EMERGENCE OF AQUACULTURE AND CHANGING AGRARIAN RELATIONS

EMERGENCE OF AQUACULTURE

AND

CHANGING AGRARIAN RELATIONS

ANEEL VB BABU J

AKANSHA PUBLISHING HOUSE
NEW DELHI

AKANSHA PUBLISHING HOUSE
R 37-B, Vani Vihar, Uttam Nagar
New Delhi-110059
Email: ektabooks@yahoo.com

Showroom
4649/21B, Ansari Road
Darya Ganj
New Delhi - 110 002
Ph. : 011-23263193/ 9811582579

Emergence of Aquaculture and Changing Agrarian Relations

First Published 2005
ISBN 81–87606–86–X

PRINTED IN INDIA

Published by M.P. Misra for Akansha Publishing House, New Delhi and Printed at Tarun Offset Press, Delhi.

Dedicated

to

MY PARENTS

Foreword

Post-Independent rural India has witnessed substantial amount of changes in the relations of production and the gap between the owners and non-owners of means of production got further widened. Thus, the changing face of Indian agriculture has a profound bearing on the socio-cultural and economic life of the people. It is important to take note of the green revolution and other similar processes, which have unleashed different forces, have diversely affected the rural livelihoods.

After 1990s the economic liberalisation and globalisation have further impacted the rural, and also urban, economies to the detriment of the lower-end farmers as well as the landless. A large proportion of these belong to the Scheduled Castes, Scheduled Tribes and other marginalised Backward Classes. The production systems have become more capitalistic and market oriented. The most significant development is that the markets—local, regional, national as well as international-have started to dictate the production and labour.

Commercialisation of agriculture thus became a major force that altered the rural face of India. This has also drastically changed the traditional agrarian relations and has now threatened the food security of the vulnerable and marginalised groups. In a way, the traditional moral economy, though has some feudal overtones, had at least ensured subsistence to the rural poor. The commercialisation of agriculture and rural economy gave rise to more impersonal, unjust and ruthless economic order. No wonder today we find more suicides of farmers, weavers, artisan and other service castes.

The forces unleashed by the liberalisation of economy and globalisation have affected the livelihoods of poor and the vulnerable sections, including women. This has further led to an increase in the migration of marginalised groups to urban places, including bigger cities. Some of these migrations have created untold miseries to the migrant groups and they have been subjected to lot of exploitation. The shift from cultivation of cereal crops and pulses to cotton, tobacco, and other non-food commercial crops subjected farmers to the vagaries of markets and market forces and thus they became vulnerable. Agriculture now turned into a gamble and farmers got lured into its attractions and trappings. These new crops needed additional capital and other inputs like seeds, fertilisers, pesticides, water, electricity, etc., supplied by the avaricious traders, industrialists and others who exploited these vulnerable farmers. Many a time if the farmer is fortunate he might recover his expenditure and make a little profit (in real terms this does not even ensure his food security). Otherwise, he will be mired into debts. This, in most of the occasions, led to the pauperisation and turned a farmer into landless migrant labour in search of green pastures elsewhere.

In the coastal regions, some of the farmers have either acquired new lands or turned their own lands into aqua farms or into pisciculture farms. Initial gains or profits have encouraged many medium, small and marginal farmers to convert their agricultural lands into prawn farms. These required huge initial investment and continuous technical support. Non-Indian firms controlled the markets as well as aqua feed companies, in most cases. When the aqua farms were affected by diseases/viruses the farmers lost successive crops and the plummeting prices for the Indian prawns became a death knell to the aqua farmers. Besides, once the farms were converted into aqua farms they became non-retrievable for cultivating agricultural crops. They also have environmental implications. The aqua farming has shifted the locale of family labour and also displaced traditional women labour force that became unemployable in the aqua farms. The lure of short-term profits has completely destroyed the age-old sustainable and dependable agriculture and livelihoods related to it. What is quite depressing is the role of the governments in promoting this shift in production gamble.

In this context, this work of Dr. Aneel is quite timely and it is based on an intensive fieldwork. It brings out the play of different forces in the uncertain shift in production systems and their ramifications for different actors and their livelihoods. His study is in a way an anthropological investigation of the otherwise field of economists. He brings out the nuances of social and cultural factors involved in the changes in the production system and its implications to the poor and deprived section who really matter. I am sure that it will open a new interest in the village studies under the changing circumstances and offer better economic and environmental policy directions for a better tomorrow to the rural poor as well as to the poor peasantry.

DR. R. SIVA PRASAD
Reader,
Department of Anthropology,
University of Hyderabad,
Hyderabad-500 028

Acknowledgements

During the course of the present study I received help from different quarters. I express sincerest thanks for all who helped, directly or indirectly, for the completion of this book which is my doctoral thesis but for this would not has been possible.

I deem it privilege to thank my research supervisor Dr. R. Siva Prasad for the continued assistance and untiring effort. He had been a continuous source of inspiration, encouragement and patiently oversaw the work till its destiny. All my teachers in the university had taught me to critically look at things, analyse and comprehend the issues in and around.

I am greatly indebted to the villagers of Godilanka who had been of help throughout my fieldwork. I am indebted for the hospitality and kindness shown to me during fieldwork. In this regard, I fail my duty if I do not acknowledge the timely help that I had received from key informants of the village.

Above all, my friends late Dr. Narayan Reddy & Ali had always been a constant source of inspiration. My associate Mohammed has extended many a kindness to me and always kept my spirit and enthusiasm high, and others who need a mention, Narendera, Nandu and Panvan from the department. Friends are always a source of inspiration and they have made me vigilant through their discussions. Keshav, Kanaka Rao, Dharma, Sudhakar (yagni), Laxmi, Narayan, Seshagiri, Bathula are a few to mention in this regard. Sashi, Sudershan, Sudhakar, Ramakrishna, Macha Venkat, Ajanta have always provided their companionship and emotional support all through the study. Praveen, Milton, Vijaybhaskar, SriSri, Eswar, Manikya Rao, Mallik, McMurthy are the 'magic people' who are not to be forgotten.

I am obliged to my mother who designed my disposition, from the cradle, to this stage who is responsible for the state of mind. My brother, Suneel, and sister-in-law; sister, Ranjani, had always been a source of encouragement. I show appreciation to my wife Roja without her active cooperation this doctoral thesis may not emerge in this form. I am grateful to Mrs. Siva Prasad for the support and hospitality during final stages of this work.

ANEEL VB BABU J

Contents

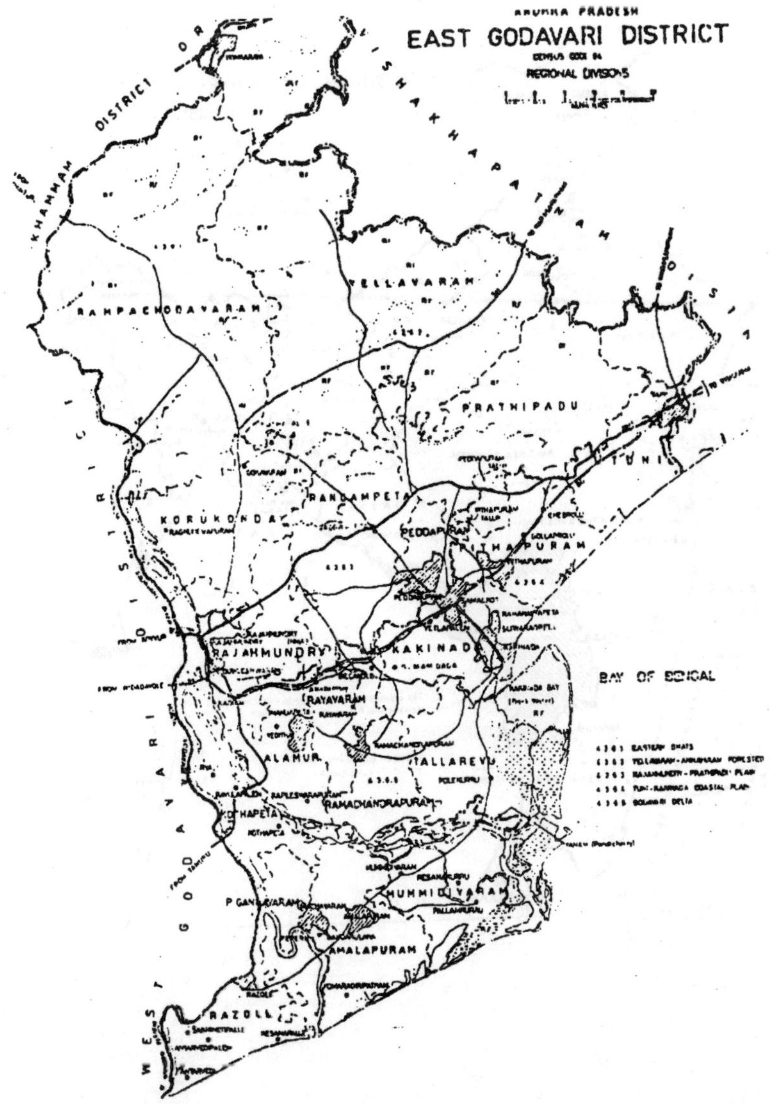

Map 1: East Godavari showing Study Village, Godilanka

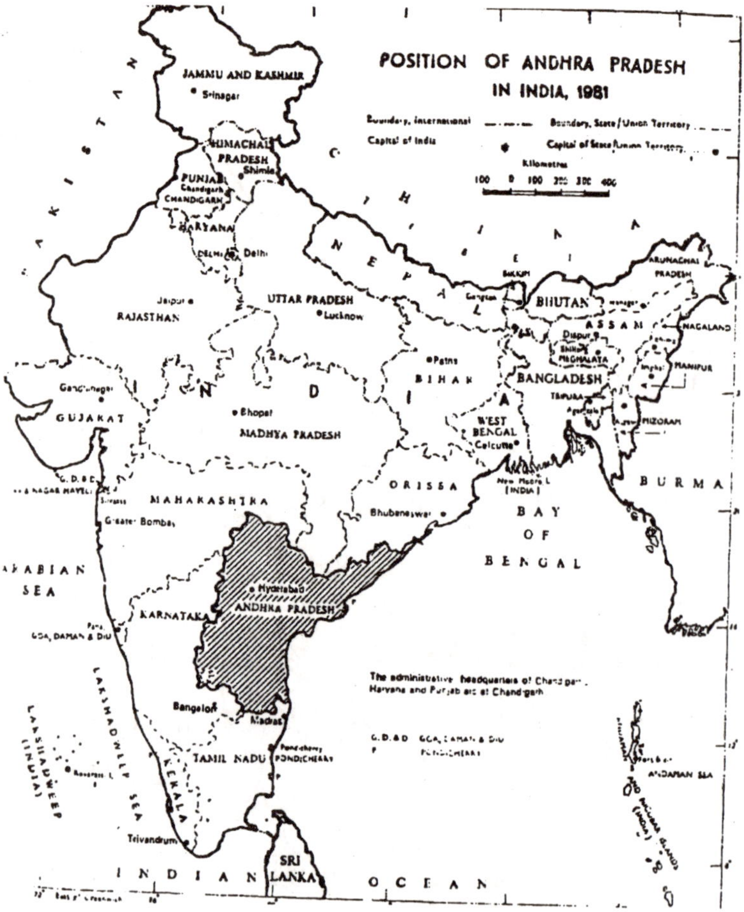

Map 2: India showing Andhra Pradesh and East Govadari

CHAPTER 1

Introduction

In a country like India, where majority of the population is dependent on agriculture, agrarian relations play an important role to understand the rural social structure. With the onset of new economic policy and globalisation the local economic systems are now connected to the global markets and are thus influenced and affected by them. The emergence of aquaculture in the coastal areas, particularly in South India, in 1990s, for direct export of prawn is a consequence of the policy of economic liberalisation.

The important thing to be noted is that aquaculture is not agriculture, but is done on agricultural lands, as the agricultural lands are converted into aqua farms. Thus, the emergence of aquaculture has its impact on the agrarian relations and thereby, it is important to study the impact of aquaculture on the agrarian relations in a village social system. Hence, the present study gains its significance, as it is useful for the academicians interested in theory building. Further, it is of value for policy makers, administrators, development workers, aid-agencies (both national and international), environmentalists and others involved in action research in understanding its impact on the populations influenced by it.

The agrarian relations in India have been altering due to many reasons. Social scientists have identified different factors such as industrialisation, modernisation, improving transport facilities, population pressure, land reforms, green revolution, new technology relating to agricultural development, irrigation, higher productivity, commercialisation of agriculture, cash cropping, new values, ideologies, politicisation of rural life, etc., that are responsible for this. The agrarian relations, in the process of time, got adapted to the changing situations in the rural society.

I

Agrarian relations are primarily class relations among various people involved in agriculture production, which, in turn, are articulated into the caste system in India (Rao, 1986: 2). As a result, the agrarian system has been considered as a part of caste stratification with upper caste owning the land, the middle caste cultivating it, either as tenants or sub-tenants, and the lower castes working in the agricultural fields as wage labourers.

Agrarian structure is differentiated into land tenure, organisation of agriculture and the influence of general economic conditions on the agrarian structure. The various aspects of agrarian relations between the landowners and tenants, on the one hand, and their relations with agricultural labour, on the other, forms the basis of integration as well as conflict in any Indian village. However, their interplay can be observed in the power relations among and within different caste groups in the village. The combination of social relations, economic relations and power structure, including mechanisms of conflict management, gives a village a basis for its organic unity.

In the study of Bisipara Bailey (1958) dealt with the economic and social change that had altered the relationships within the village economy as a result of land becoming a marketable commodity. The study disclosed the changes that have come about in kinship relations of an agrarian community on account of changes in its economic and political structure. He maintains that, "Organic solidarity arising from the division of labour is something seldom overlooked in studies of the economic function of caste. ... But there is another aspect of the system of caste-groups in a village community. Certainly in Bisipara, and probably in village communities elsewhere, the ranking system of caste groups was validated not only by ritual and social usage, but also by differential control over the productive resources. ... This system still survives to-day in the tithe payments ... applied to ... 'village servants' (and) ... to the 'client' castes..." (1958: 266).

He further avers that, "In seeking to improve his position in the ritual system of rank in his community, the rich man cannot throw off his poorer caste-fellows: he must carry them along with him. There are clear reasons for this. The new economy has done nothing to break down caste endogamy. Therefore a man must find wife from among his own group. He depends on the rest of his caste for help in the rites of marriage and death and at numerous others crises in the course of his life. With such ties as this, the most that the wealthy within a caste-group can do to lighten their upward passage by jettisoning poorer caste-fellows is to form a sub-caste" (ibid: 270). He noticed in Bisipara that, "... the local structure is undergoing a fundamental change. It is losing its political-economic function. This function rested on the fact that productive resources were allotted to caste, according to caste, and control of these resources was organised by a localised structure of caste-groups. In one case the structure has adjusted itself successfully to the new economy. But in the other cases it has failed. The new economy is widespread: caste-group organisation is appropriate to a localised economy. The new economy demands social mobility"...(1958: 275).

Gupta (1963) opines that agrarian relations are economic relations amongst people in the rural society as they characterise the factors of production in the organisation of agriculture and in the distribution of income. Viewed differently, agrarian relations are also social relations of production. The terms 'agrarian structure' and 'agrarian relations' are used interchangeably by different scholars. In order to understand the dynamics of agrarian relations, it is essential to examine the interplay of different socio-political and economic factors that operate in rural settings as part of a village social system.

The important feature of village economy, in India, before the coming of British was that the peasants were broadly in a subsistent economy and were self-sufficient. They were under the authority of a local prince or chieftain, who in turn was under an overlord, whether Raja or a Moslem Nawab. The peasants conceded the authority of the local chieftains by paying an annual share of their produces. This varied between one-sixth and sometimes to one-half of the produce. The chieftains in turn paid a part of it to the overlords as tribute. These agrarian relations were concomitant with social and economic relations and were manoeuvred by prevalent customary patterns and conventions (Thorner and Thorner, 1962: 51).

Thus, the basic relations in agriculture were deep-rooted in custom. Driven by custom, the farmer used to till the soil and reap the harvest.

This continued as long as he rendered his services to the immediate local leaders or representative of the overlord. When this right was abused and these local leaders became harsh, then the peasantry moved to places where it was conducive. During that period, since land was abundant and labour was scarce, the local chieftain and feudal lord were careful in exercising their power so that they were not alienated from the villagers (Thorner and Thorner, 1962: 52).

The relationships in the Indian villages were developed over a period of time and they were based on a ritual ranking (Jha, 1977: 21). Sedentary life and stable agriculture led to sufficient production of food grains, which was appropriated by kings and nobility in the form of taxes.

During the Mughal period, the village was recognised as a unit of administration and this remained unchanged through ages. Some scholars held the view that private landholding existed during the Mughal period. The possession of private land has come into existence during the Mughal period (Naqvi, 1975: 78; and Gough, 1980: 336). Gough observed that there was a mixture of private and collective ownership in the beginning and in the later stages there was an increasing tendency towards private landholding by peasants. Even in south India "both types of ownership of land might have existed at some places, the agrarian systems might have been in a transient stage" (Rao, 1986: 17).

Great prestige was attached to the possession of land. Irrespective of the occupational background, all classes of people, excluding the ex-untouchables, possessed agricultural land. All these categories jointly shared the proceeds in agriculture (Sastri, 1966: 327). The ex-untouchables, though were part of the agricultural operations, were not entitled to own any agricultural land.

After the advent of the British, the concept of private property became prevalent, as land became a commodity that could be sold, mortgaged or purchased. The self-sufficient village communities formed the basis of social security and social control. The British considered these as an obstacle for economic development. In order to overcome this, they introduced new land revenue policies, which dismantled the isolation of village communities. Thus, the village economy became part of wider market linked transactions.

The introduction of Zamindari[1] system strengthened the hold of landlords over the tenants. This has led to the concentration of land in the hands of a few and dispossession of land from the others, who became landless. Mukherjee (1975: 21-30) opines that landlords and supervisory farmers formed the highest rung and sharecroppers and agricultural labourers formed the lower levels of agrarian hierarchy. The artisans and service castes depended for work on cultivators. Thus, the entire organisation in agriculture was of subsistence character. The new revenue system introduced by the British had disrupted the organic unity of the village communities. Agriculture as a system acquired a commodity value.

The introduction of permanent settlement[2] had created the concept of private property and this had a debilitating effect on the communities. This system of permanent settlement had brought in greater inequalities among the communities. On one hand, it had prevented the peasants from the ownership of land and they were condescended to the level of sharecroppers and labourers and, on the other, a new category of landlords came into existence. These were mostly absentee landlords and exploited the peasantry in different forms (feudal/semi-feudal). Thus, with this system the British have altered the economy to commodity production and collected high revenues, which were invested by them elsewhere. They were successful in introducing a production more or less based on feudalistic lines (Gupta, 1963; Dutt 1970: 80).

A different type of land revenue, the ryotwari[3] settlement, was introduced in other areas where the small owners of land were considered as instruments for agricultural progress (Gupta, 1963: 2). In this, the right to land depended on annual payments to the State revenue. Thus the rights and privileges of different classes and categories of landowners were dependent on different sets of conditions to possession of land. These rights and privileges of various categories to a large extent were governed by their caste rank. Thus, the British succeeded in introducing landlordism, a semi-feudal tenure, and succeeded in creating a new set of relationships. A hierarchy of tenants on one side and the landlords on the other, while the bulk of masses who toiled on the land were not given due significance (Singh, 1975: 61).

During the pre-independence period two distinct views were held for the economic development of India. The institutional approach viewed that changing the semi-feudal agrarian relations as essential and necessary. The other emphasised on the need for technological progress for bringing changes in the agrarian relations. The above two views are not mutually exclusive though they appear divergent. The former tried to alter the land system, where the landlord never worked on land but got his share for indulgence in managing the labourers who worked on it. The later, was to convert the mass of tenants into owners so as to eliminate the landlords, who were supervisory in nature (Joshi, 1967: 447-8).

The British considered Indian agriculture as backward and this was due to its technological backwardness. They never tried to relate it to the agrarian framework. Their efforts involved mostly in the development of irrigation network, etc., which were of technological nature, to increase production and undermined the importance of changes in institutional framework. Indian nationalists also held a similar view and they gave little importance to understand and analyse the landlord-tenant relations. However, a few opined that the institutional framework of agriculture was the main obstacle for economic development and welfare. They held that a change in the institutional framework would bring progress and development and mitigate the problems in Indian agriculture (Joshi, 1967).

II

The agrarian social structure and the inequalities of Indian rural society were viewed mainly either in caste or class perspective. The mutual dependence of landowners and those who worked on it was the predominant and important characteristic feature of the agrarian relations in India. These relations were viewed in terms of inter-caste relations, either as *jajmani* or patron-client relations (Wiser 1958; Leach 1960; Dumont 1970 and Bailey 1958).

Agrarian relations in the early part of the twentieth century were studied and analysed by the British administrators, travellers and other researchers. They viewed the village social structure and agrarian relations as inter-caste relations. These studies provide an understanding of relationships among various castes in an agrarian social structure.

The provincial and imperial Gazetteers of India (1908) and Census of India (1901) provide an account of Indian society in the second half of the 19th century. They contained mostly the revenue settlements, an account of land, people, and social organisation. In the1901 Census of India 'social precedence" was recognised on the basis of caste ranking, which was elicited from the natives in a given region. The caste relations were regarded as most significant in their studies.

The early systematic studies about village communities and their life styles were made by Munroe, Metcalfe, Maine, and Baden-Powell. They viewed that agrarian relations have characterised the Indian village as a 'closed' and 'isolated' system. Since 1950s there was an increase in the studies of rural agrarian systems. Particularly, 1955 was a landmark in the studies relating to agrarian relations among village communities and the most prominent studies published were of Dube's (1955) 'Indian Village', Srinivas' (1955) 'India's Villages', Majumdar's (1955) 'Rural Profiles', and McKim Marriott's (1955) 'Village India'. These studies covered a vast range of data and experiences from different parts of the country. They particularly dealt with the lifestyle of the landowners as well as those who lived on land as part of a village system. From then on the agrarian relations, economic structure, political system, social relations and the way of life among them as part of social structure became important aspects of study. Thus the emphasis of studying village social structure as part of relations continued to be a central theme of anthropological and sociological research.

Following this, the Directorate of Census of India had undertaken studies in villages of every state. The main aim of these studies was to obtain information related to village statistics, a brief account of social structure and to understand 'the winds of change' in the villages. Thus, a large number of village surveys and reports on tribal and rural settlements were compiled. Most of the reports had details relating to castes, its distribution and their intricacies, ownership of landholdings, indebtedness, material culture, etc., (Bose, 1984: 7). Thus, during this period the emphasis of studies in understanding the agrarian social structure throughout out the country has moved from single village studies (holistic) to comparative studies of villages (Dube, 1958; Epstein, 1962; Ishwaran, 1966; Mukherji, 1957; Singh, 1969). In these studies the emphasis had shifted to large-scale surveys at macro-level using research methods other than fieldwork as in micro-level studies (Srinivas, 1975; Saberwal 1972). Thus the studies during this period had extended beyond the village level.

Agrarian relations in a village were predominantly studied as *jajmani* relations. The *Jajmani* system[4] as an analytical framework was originally used in the anthropological literature after Wiser[5]. This was viewed as a patron-client relation with mutual obligations of work and rewards, which is socially determined. The payments were through customary mode based on local conventions of inter-caste relations.

Many studies were conducted on agrarian relations as inter-caste relations (Wiser, 1936; Omvedt, 1980).

The *jajmani* system was regarded as pan-India phenomena (Dumont and Pocock, 1958: 43-44). The traditional system protected the villagers against subsistence crises, such as food shortages, droughts, floods, epidemics and calamities. Scott (1976) referred to the 'safety first' principle that would explain the behaviour of farmers in a pre-capitalist society, which implies that choices tend to reflect risk-avoidance instead of profit maximisation. The most important method to reduce subsistence crises was to share these risks with other members of the village community.

Jajmani payments were also paid when the crops failed. This system enabled the upper castes to maintain ritual purity and to display their prestige and status as the Kamin had a subordinate role. Changes in this system was due to the removal of autocratic power from villages, increased opportunities for mobility, development of communication, technological advancement, changing values and norms, legislative enactment to follow any occupation and enforcement of authority of State (Bose and Jodha, 1965: 119-123).

The possession of land ownership that yields power dictates the nature of payment in *jajmani* relationships. This type of system, which is considered exploitative, based on power relations, was also observed in the studies of Majumdar (1955: 211) and Reddy (1955: 137). Increase of population and decrease of landholdings coupled with the growing demands of lower castes have been considered responsible for the decline of the *jajmani* relationships. The weakening of *jajmani* system in the agrarian system had increased inter-caste tensions (Lewis, 1958: 79-84).

The introduction of commercial crops/cash crops altered the nature of *jajmani* relations in the Indian villages. Harper (1959) explains this in the study of a village in Malnad where he observed the prevalence of both *jajmani* and Malnad[6] systems. In *jajmani* system the relationships are hereditary and relatively permanent and payment is in grain, which is observed with subsistence crops. On the contrary, in Malnad system they are contracted and temporal in nature. The payments in the later are monetary payments and the labour attains bargaining power because of cash crops. While dealing with the Andean peasant village, Morales (1986) observed a similar trend between traditional and modern economic systems.

Jajmani system was not antagonistic to market forces and morality *per se*, but, was influenced by factors such as increase in population pressure, labour migration, land colonisation and increase in employment opportunities (Commander, 1983: 311).

Kumari and Ranga Rao (1993) dealt with the nature of *jajmani* relations and the pattern of occupational mobility, in Andhra Pradesh, which depended on the practice and overall functioning of the village system. Kamins, in addition to traditional occupations, followed different occupations with a degree of 'functional interchangeability' without violating the principle of purity-pollution. The *jajmani* relations were regarded as a hurdle to change but the benefits of development seem to be breaking the traditional relations. The nature, extent and direction of such change are essentially a function of the level of development and the opportunity structure available in the village and the region. Benson in his study of a Telangana village emphasised on understanding '*jajmani* systems from a regional point of view would provide greater insights not only to village economic systems but to intercaste and intervillage relationships in general' (1996: 249).

In a critical understanding of *jajmani* relations, Fuller, quoting Bayly (1983), observed that, "Some recent historical work ... challenges more powerfully the assumption that villages largely remained outside the domain of monetary exchange and the market economy. Bayly . . . criticises the assumption that the pre-colonial state was 'little more than a machine for extracting tribute from entrenched local communities, and that surface political changes hardly affected the conduct of rural life' . . . and that the peasant farm 'made its own production decisions in isolation from the interface of outside elites, excepting demands from rent or revenue' . . . the actions of the state and non-rural elites could directly draw farmers and other villagers into a wider economic system which displayed the unmistakable sign of market-oriented development. . . . Nevertheless, it still appears reasonable to conclude that the extent to which villages in pre-colonial India were integrated into the wider economy and were penetrated by a market mechanism mediate by monetary exchange has probably tended to be underestimated. That 'customary', non-market exchange in kind dominated each village's economic system is clearly an assumption, present in many anthropological descriptions, which is incompatible with the weight of historical evidence" (1989: 46-47). According to him, "The demand for cash revenue was a major factor to integrate villages into a wider monetised economy" (ibid.: 45).

Lerche (1993) observes that *jajmani* system as such does not exist as an all-India phenomenon and he does not accept the two pronged relationships, which were talked mostly in the concept of purity and pollution. According to him, *jajmani* relations are perceived by people distinctly as three different relationships. He accepts the King-Brahmin model proposed by Pocock. These three relationships are between landowner and Brahmin, landowner and servicing castes and landowner and servants. He agrees with Miller (1986) and argues that regional specificities have to be taken into consideration.

Interestingly, Karanth in a study of a Karnataka village, Rajapura, observes that, "It is necessary to view indentured and agricultural labourers from the perspective of caste, since caste membership is a vital element for participation in the *jajmani* relationship either as a patron or as a specialist. Members of almost any caste could be indentured or agricultural labourers (with the exception of ritually high castes, in the past). . . . Mere continuity of indentured relations or landowner relations over a long period does not necessarily amount to *jajmani* relations It would be more appropriate to consider them as 'patron-client relations', which may include *jajmani* and other relations of mutual interdependence" (Karanth, 1987: 2217). He further argues that, "the institution of *jajmani* relations has adapted itself to suit the changing needs of the people both in the past and in recent years. . . . An examination of *jajmani* relations in the context of a new technology, namely sericulture, shows that the latter has not significantly affected one of the traditional forms of rural institutions" (ibid.: 2222).

Breman (1974) viewed that in pre-capitalist village society several social systems existed that was based on mutual dependency of the village. Patronage in such a system is in exchange for their labour, their political loyalty and general servitude, the clients received protection and various forms of material support from their masters. The *jajmani* system is another example of a traditional village level system regulating the exchanges of goods and services on non-cash basis.

Some scholars considered *jajmani* system to be an aspect of caste (Bougle, 1971; Hocart, 1950; and Dumont, 1980: 35). Others viewed it as an exploitative and unequal system of relations. Parry (1979) studied the interrelationships in Kangra[7], through patron-client relations, which fall in the same mould with a little change in tone and content. Thus, there was a disagreement between those who conceptualised the agrarian system as a harmonious interdependence in which there were privileges for the low caste groups (Wiser, 1936 and Leach, 1960) and those who emphasised *jajmani* relations as asymmetric power relations between patrons and clients (Lewis, 1958; Breman, 1974; Miller, 1986; Meillassoux, 1973; and Orenstein, 1962) and they insist that 'any integration possible is of a coercive nature' (Beidelman, 1959: 68).

In a recent study Karanth (1995) dealt with the nature of emerging agrarian relations in the context of development programmes, such as land reforms and sericulture[8]. He regards land, technology and network relations in the agrarian social hierarchy as core elements of agrarian social structure. He analysed the nature of inter-caste relations among mulberry cultivators within the framework of existing agrarian relations and nature of labour arrangements. He opined that the break up of *jajmani* relations was due to monetisation. He observed that impersonal and bargaining relationships came into existence because of sericulture.

While *jajmani* or patron-client relations are regarded as reflecting agrarian relations in village India, caste, kinship and locality are closely related to social organisation of agriculture Kumar, 1965: 190 and Béteille, 1974: 56). The identification of an individual among peasant societies in India is through extended family, *jati* and affiliation of an individual to the village (Sachchidananda, 1988).

AC Mayer (1960) studied the agrarian social system in relation to caste and kinship in a village community and also in its region. His study shows that caste membership is significant for relations with others in a given social system and membership in a sub-caste for activities in a caste. He argues that, "Economic factors lie behind the patterning of the various kin-groups making up the subcaste, as well as the degree to which members keep in touch with each other. For the scattering of many kin-groups, which has helped to produce a weak unilineal structure, has been partly for economic reasons. . . . At the present time, movement from villages to towns is increasing since land is scarce. . . . Both past and present factors make for a situation in which kinsmen are scattered. And under such conditions the economic ability to invite kin to weddings, mortuary feasts, etc., is important, as is the inclination to do so. It would indeed be difficult to resist kinship obligations if one's kin were in the same or adjacent villages. . . . But on these gatherings rests the continuance of the kin tie, and people identify kin ties with invitations and so with money" (1960: 273). Siverstein (1963) in a similar type of study observed that the caste system is important for analysis of political structure in the rural social situation.

III

The understanding of agriculture as a system of production on class lines, in a rural society, was emphasised by some scholars. Earlier, the production in agriculture was of subsistence nature and the emphasis was on maintaining the relations with in a village, between the landowners and their clients. The use of land as a source of investment has altered the production type as well as the relations. Viewing the agrarian relations through class lines was done after the writings of Marx.

Marx views that generation of surplus value is the core to capitalist mode of production (Marx 1954: 618). Lenin understood this type of production from the historical experiences of development of rural economy in the former Soviet Union. He emphasised this on three major distinguishing features in agriculture, viz., 1) the production for market; 2) improvement in technique and method of cultivation and 3) the use of hired labour paid in cash (Lenin 1967: 318-23). Indian agriculture was examined in the Lenin's paradigm and was recognised that there was also an emergence of capitalist relations in Indian agriculture (Sen 1962: 174-175). Sen argues that even after the production had surplus, the relations in Indian agriculture were mostly of semi-feudalistic in nature.

There has been a definite trend towards the growth of agriculture in India after independence, leading to the emergence of two distinct classes viz., the capitalist farmers, and agricultural labourers. The development of capitalism has its origins in colonial period and Utsa Patnaik (1971a, 1971b, 1972a, 1972b and 1973) and Chattopadhayay (1972a and 1972b) proclaim that commodity production, capital investment and employment of wage labour in agriculture are products of colonial exploitation. On the contrary, Gupta (1969) argues that the evolution of capitalist development in agriculture was a post-Independence scenario.

The capitalist landowner employs labour as worker/wage labourer in an industrial enterprise. For him maximising profit is the only motive. As a consequence of this the established relationship for ages between the landlord and the people associated with the agrarian set-up appeared to be undermined or redundant. The servicing and the agriculture-associated groups' (such as 'artisans') nexus is mostly abandoned, except in case of big and large landowners, and a new category of people, landless agricultural labour, appeared in the agrarian scene.

A number of studies dealing with the emerging capitalist classes in the rural society were undertaken to know about the mode of production in India (Kumar, 1963; Shanin, 1977; Prasad, 1980; and Blair, 1980). As a consequence of the capitalist mode of production, there was a category of landowners that owned land and hired labour to work in their fields for wages. In between, there was a category of workers, the tenants, who worked on the land which was not their own but leased the land from the land owners for rent. Thus, the agricultural labourers as a special category emerged in the agrarian scene. The growth, type (attached or free), and wages of agricultural labour in India were widely studied (Patel, 1952; Krishnamurty, 1972; Dhar, 1973; Gosh, 1948; Bandhopadhyay, 1977; Thorner and Thorner, 1962; Breman, 1974; Bardhan and Rudra, 1980; Alexander, 1973, 1978, 1978a; Krishnaji, 1971; Bardhan, 1980; and Mencher, 1980).

A decade after independence, Indian agriculture underwent land reforms as a part of the programme to revive the structure of agriculture. During 1950s, the Government had introduced land ceiling act and land reforms. The main purpose of the programme was to eliminate the intermediaries and pass on the land to the actual tiller of land, who would be paying taxes directly to the State rather than paying rent to the landowner. This intention of reforms was to bring about changes in the agrarian structure and relations. Land reforms were made part of the development programmes of the State and their role in the reduction of poverty, unemployment and inequalities was almost found to be negligible (Oomen, 1971; Mukhopadyaya, 1975; and Page, 1975). It is prevalent even today that manual labour is regarded as degrading and is to be performed by inferiors, who belonged to the lower castes and classes.

The relationship between land ceilings and agricultural land holdings, and social mobility reveals that they had affected upward mobility of middle farmers (Viswanadham, 1975). Viswanadham and Swarna Kumari's (1980) study in coastal Andhra showed that the concentration of land had decreased. They had commented that besides land reforms other factors such as demographic factors had also contributed to it. The studies made by scholars like Tharant (1962), Sinha (1975), Pande (1982), Sen (1982), Ranga Rao (1980) and Ballal (1983) had expressed that division of landholdings between brothers and relatives were not the intended goal of the programme. This did not meet the very primary objective of land reforms and land remained still with the same landowners from the upper caste/classes. As a result, there was an exodus of medium and small farmers from rural to urban areas. It does not mean that land reforms had become obsolete. They did alter, to some extent, the structure and relations in agriculture.

To attain self-sufficiency in producing food grains to meet the demand of the growing population and crop failures in the preceding years, a programme with new agricultural strategy was introduced. This was popularly known as the 'Green Revolution'. Green revolution had its impact on the agrarian relations and in turn on the agrarian structure. It generated growth and prosperity. However, it was an indirect cause for widening the gap between the rich and poor in the rural areas. It has produced a feeling among the agricultural labour that they were not getting an adequate share in the benefits of increased agricultural production and the cultivators appropriate these benefits. This fact has quickened the process of economic polarisation in the rural areas (Alexander, 1971). This has led to increasing social antagonism between landowners and tenants or landowners and landless agricultural labour (Béteille, 1974). The green revolution to a large extent accelerated the production of food grains and altered the agrarian structure and relation in general, and the nature of production relations in particular.

Seed-fertiliser innovations, which were introduced as part of green revolution, have encouraged spread of farm mechanisation (Parthasarathy and Abraham, 1974). Technological changes in agriculture in the wake of green revolution have their consequences on village social structure/society. After the introduction of green revolution, irrigated areas needed more labour during all stages of paddy cultivation. Consequently, the landowners had to depend on the labourers primarily. This had increased the bargaining capacity of labour. Their wages were increased, as multiple cropping needed quick work. Mechanical devices were introduced to help the farmers in faster seedbed preparation and other activities as demanded by multiple cropping and HYV with irrigation facilities.

The green revolution in principle was considered scale neutral as it raised yield and income for both small and large-scale farmers. It was argued that large farmers were the main adopters of the new technology and the smaller farmers were either unaffected or badly effected. Green revolution resulted in lower productive price, higher input price, increased rents, tenants got evicted of the land and larger farmers increased their landholding by purchasing small farms, thus small and marginal farmers were turned into landless. It led to rapid inequalities in income and asset distribution; and worsening of poverty in the areas affected by green revolution (Griffith, 1972, 1974; Franckel, 1976; Harris, 1977; Hewilt de Alacantara, 1976; ILO, 1977; Pears, 1980; Dhanagare, 1982; Byres, 1981; Minch, 1978, Parthasarathy, 1970, 1991; Epstein, 1978; Dasgupta, 1972; Mohanty, 1999).

In a study of rural transformation Somayajulu and Mahanth (1984) compared, a wet and a dry village of AP, in relation to cropping pattern, mechanisation, infrastructure facilities, financial institutions, social services, life-styles and lord-labour relationship and political awareness. They observed that, "traditional feudal and semi-feudal modes of production, relation between the rich farmers and landless agricultural labourers, servants etc., still persist". The green revolution has increased social inequality in the wet village.

Others argued that green revolution improved the economic conditions of poor peasants and landless agricultural labour. The new incomes had improved their standards of living. While the workdays have increased, their bargaining capacity has also increased, and, further, it led to a reduction in inequalities (Byln, 1983; Hazal *et.al*., 1991; Ahluwalia, 1978; Lipton and Longhunt, 1989; Miller, 1978; Rutten, 1977; Shergill and Singh, 1995).

There is a third view about the green revolution. In addition to the benefits to the landholders, it also had improved the conditions of the small farmers and landless agricultural labourers. Modernising of agriculture as a package had combined effects for both owners and labourers (Bhalla and Chandha, 1983; Naidu, 1997; Athreya *et al*, 1990; Harris, 1991; Agarwal, 1983; Dak, 1989; Arora *et al*, 1991; and Oomen, 1989). Most of the studies confined their analysis to the micro level, concentrating on the individual aspects of modernisation. Since the natural and social environments in South Asia are not similar, the affects of new technology on different communities were varied (Farmer, 1986).

Large-scale farmers especially dominated the green revolution, which increased agricultural growth, in India. Farming was accompanied by rapid growth of mechanisation. This needed large amount of capital and also resulted in large quantities of surplus production (Rudra 1969, 1969). Thorner (1969) points out that there were a few pockets of capitalist farmers in India, in certain regions, even before independence. Their emergence as a significant group in every state is one of the facts that technology, which is a result of industrial revolution, has changed the face of post colonial India. Rudra (1970, 1971) points out that the changes talked about by Thorner are qualitative rather than quantitative in nature. These changes have taken place within the framework of old production relations and class structure had remained unaltered.

IV

Change is an ever-present feature in human civilisation. It can be said that the tendency to change is fundamental to society and it renews the existing structures, practices and customs. Every structure in general, is said to be having two kinds of forces—one that promotes changes and the other, which strives to maintain *status quo*. The former represents the stimulants of action promoting change and the latter change-inhibiting factors or barriers (Foster, 1973).

There has been a growing interest among the anthropologists in the progress and impact of economic development in rural India. This is evident from the studies of Marriott (1955), Bailey (1957), Opler and Singh (1957), Dube (1958) and Epstein (1962), who had dealt with change in rural India. Opler and Singh (ibid.) argued that change operated at different levels—local, regional and national.

The agents of change in the Indian agrarian system, which has contributed to the economic development, had brought about impact and change in the agrarian social structure. They may be structural or positional changes. The notable to mention are improvement of irrigation facilities, introduction of Land reforms, introduction of new technology, commercial and cash crops, green revolution, mechanisation of agriculture, etc.

Irrigation played, and still continues to play, a very significant role in development of agriculture and rural areas. Studies on the impact of irrigation in different parts of the country were both synchronic and diachronic covering macro (regional and local level) and micro levels (village level). The main thrust of these studies was to bring out the impact of irrigation on various aspects of village, whether at the level of household or village, such as farm economy, general economic conditions, employment, and social change.

Epstein's (1962: 179-193) study is an important study in this regard. Her's is a comparative study of wet and dry villages. Irrigation had brought large areas of land into cultivation and, consequently, the production has increased. In the wet village, which continued to be agricultural, the traditional relations in the village social system were reinforced, economic status depended on the size of the estate, number of heirs and, importantly, individual desire and motive. In contradistinction, in the dry village the economy is diversified and the division between progressive and conservatives became distinct and a third force, who worked outside the village but associated with the village, grew into an influential group within the village. Land was divided, non-agricultural income supplemented and increased the income of the landless and new social relations have emerged among the villagers. In both the villages, new rich became the contenders for prestige and power bride wealth changed to dowry.

The character and pace of the transformation of Indian agriculture after independence has been the subject of research. Gorter (1989) maintains that in early 60's agriculture was already commercialised and wage-labour used on a permanent basis. The construction of large-scale canal irrigation system has converted them into full-fledged rural capitalist entrepreneurs. As a result of irrigation, cropping pattern has changed. The traditional cash crops disappeared and in most of the places sugarcane, which was more profitable, replaced the traditional crops. Paddy and area under vegetables had increased. Crops were no longer grown for house consumption. Permanent wage labour came into force and more migrant labourers were hired. Canal irrigation profited big landowners, while small and marginal farmers had to rent or sell their land and it transformed the big landholders from commercial farmers to full-fledged rural capitalist entrepreneurs. The big farmers had not invested on sophisticated technology but retained the local agricultural labour, which was more related to status considerations rather than economic calculations. (Gorter, 1989).

A wide range of topics relating to the agrarian social structure was studied under the impact of irrigation. They have covered on double cropping, crop intensity, cropping pattern, yields, employment, cultivation practices, socio-economic conditions, standard of living, etc. The introduction of cash crops, in this context, had changed the nature of relationships among the different categories/classes. The relations were more monetised and showed an inclination towards contractual rather than the patron-client relations. It was observed that there was an increase in employment of labour and payment of wages in cash, shift in the occupational pattern, wherein, emphasis was towards agriculture as main occupation and a majority of artisans had joined agriculture (Krishna Murthy, 1956; Singh and Singh, 1962; Hanumantha Rao, 1965; Jha, 1987; Breman 1984; Sachidananda, 1972; Khan and Tripathy, 1972; Prakasa Rao, 1976; Rajpurohit, 1977; Aziz, 1977; Pandey, 1979; Alexander, 1979; Kapp, 1959 Misra, 1979; Goldman, 1982; Patil, 1980; Rana, 1980; Vasudevarao, 1987; and Satyanarayana, 1995).

From the above studies it could be understood that irrigation increased incomes of both the owner-farmers as well as the labourers. However this was beneficial to the lands which were irrigated and not for all. This had further led to the formation of different agrarian classes and the traditional work activity (secular or religious), which was cooperative, has almost disappeared (Poffenberger and Zurbuchen 1980).

Surinder Jetley (1977), commenting on modernisation of Indian peasants, felt that the joint family and caste system was highly conducive to imbibe innovations and they act as levers for entrepreneurial activity. The high castes are ritually as well as economically powerful but the peasants of lower castes are numerically more. Developmental programmes such as land reforms, politicisation and principles of democracy proved to be beneficial not only for the upper caste, but also for the numerically dominant peasant castes. In villages which are more developed, caste and class divisions almost merge completely. On the contrary in the less developed villages there is a weak relationship between the level of living and caste, and social participation.

Technical innovations and farm mechanisation has brought about general change in economy and society also. Farming became less arduous and increased productivity has led to increase in production (Schiller, 1959; Rao and Singh, 1964; Saini, 1964 and RaghavaRao,1979: 357). Subsistence production was substituted by production for the market, which in turn improved the standards of living. Family relationships and traditional role patterns are also influenced by mechanisation. Old women and children withdrew from active farm work (Singh and Singh, 1970). Youth replaced the old and also assumed the leadership (Sisodia and Sharma, 1972) and children now continue studies for longer period and thereby delay their entry into the work force (Pressler and Willes, 1976). With the elimination of drudgery in farm works, educated people found agriculture more interesting and this had affected their social and community participation. In the political organisation, leadership and political participation has changed. It was transferred to the young from the traditional categories (Kaushal, 1971; Sisodia and Sharma, 1972). Individual and family relations with in and outside the community had expanded (Solomon, 1960 and Slocum, 1962). Mechanisation has also led to increase in cropping intensity (Okhawer, 1961; Singh, 1970; Chauhan and Verma, 1971; Kahlon, 1972; and Motilal, 1973). Crop yields, interestingly, were found to be neutral to tractor use (Vashishta, 1975; Hanumantha Rao, 1975; Rudra, 1970; Agarwal, 1982).

Change in technology and cropping pattern had its impact on agrarian relations, quite often they altered the relationships between, and among, those who were part of the system. Further change in technology needed a change in the skills of the labourers. This has increased their work, which, in turn, increased their incomes and improved their living standards (Page, 1975). New technology and commercial cultivation led to re-interpretation of traditional social relationships (John, 1975). Parthasarathy and Prasad (1975) have observed that new technology has influenced the agrarian relations positively. The influence of technological change in west Godavari, Andhra Pradesh, resulted in surplus production in agriculture and it 'slowly disintegrated the peasants' (Parthasarathy and Pothana, 1983: A 96).

The impact of new technology and commercialisation of agriculture has resulted in the processes of differentiation among peasant households more on economic lines. This change in the rural economy is not only confined to large landowners but has also reached middle and small landowners. Mechanisation has increased the gap between the rich and the poor and created inequalities among the 'owners and non-owners'

(RaghavaRao, 1979: 354). 'Tube wells and tractors' are the new sources of power for the rural rich (Siddiqui, 1999).

Agricultural modernisation after independence, in addition to those already present, has created further inequalities. Though it has been beneficial to all, was to a larger level it favoured the landed and transformed them into rich class at the cost of small and landless labourers. Mohanty opined that, "inequality further breeds inequality unless checked" (Mohanty, 1999: A59). Barraclough (1974) viewed that in order to give advantage of developmental programmes to the low-income rural classes greater attention has to be paid towards analysing the social structure and political processes. He regards development not just as economics or sociology or technology but history and the past thus has a bearing on the process of development.

However, others (Billings, 1970; NCAER, 1980; Singh and Singh, 1970; Soltani, 1973; Singh, 1976; Motilal, 1973; Singh, 1970; General and Kahlon, 1977; and Agarwal, 1983) held that the social implications of the replacement of work by machines (tractor), which was earlier done by men and animals, was loss of both familial as well as hired labour. Further, effects of mechanisation at familial and community level were also found to be adverse in their impact (Kawamoto, 1972; Naidu and Rao, 1977).

V

Agrarian structure in India encloses a world of its own, since it neither involves large estates nor a peasant system (Warriner, 1969). This system is highly influenced and directed by the principle of caste, i.e., who would work for whom. The complex pattern of inter-caste relations, which had been set by this dimension of the stratification, has affected the agrarian social structure in various ways (Ramaunni, 1975: 223). In this context, Nirmal Sengupta (1979: 89) has observed that previously the zamindar-tenant economic relations were also in consonance with the upper caste backward caste social relations.

Social division of labour on the basis of the caste is the basic feature of agrarian society in India. The cultivating peasant castes and depressed and ex-untouchable castes, standing at the bottom of the social ladder, constantly swelled the ranks of tenants, sharecroppers, and farm servants (Joshi, 1969: 446; Mukhopadhyaya, 1980: 43; Etienne, 1968: 62; Aziz, 1980: 21-22; Shah 1969: 287 and Ballal, 1974: 85).

Generally, landowners are from higher castes (Mehta, 1980: 86) and it is this, which gives them higher status and political power in a village. In the traditional society, the higher castes had normally all-important rights, which in a way infringed on the rights of other inferior tenure holders (Varghese, 1970: 16). However, considerations of prestige as well as caste would, therefore, prevent landowners from actually working on their land, thus greatly increasing the demand for labour (Srinivas, 1966: 42).

Krishna (1980: 58) observed that the landless workers, who make-up the bottom layer of the rural agrarian society, were living for several centuries at the mercy of an exploitative system based on caste, which is operated by a combination of 'coercion and intimidation'. Even after independence, Indian agrarian scenario remained the same, the lowest castes are born to labour with their hands, and high caste to enjoy the fruits of their labour. The poor landless labourers, who were from the lower castes, were at the mercy of the landlords (Chandra, 1976: 3 and Alexander, 1973: 1557). Under these conditions, where land and caste governs status, economic power and political influence, labourers are at a disadvantage (Ballal, 1974: 85).

Marxian scholars on the other hand argued for the appropriateness of class in the study of agrarian relations. Agrarian structure was viewed as a combination of landowners and agricultural labour. In this context, structure of the agrarian classes was found to be fit and the caste system was understood basically as the manifestation of class (Desai, 1948; Gough, 1960; Millassoux, 1973; Ranadive, 1979 and Namboodripad, 1979). Similar type of analysis has also been reflected in the recent works on caste and class (Klass, 1980 and Omvedt, 1982).

André Bétellie (1974: 157) has noted that there were two kinds of relationship between caste and agrarian structure. They are: 1) the surface relationship, which is revealed by the fact that the landowners belong predominantly to the upper castes, 2) the deeper relationship in which the hierarchical values of castes sustained and legitimised the unequal relations among landowner, tenants, and agricultural labourers. In Indian social history, caste, being a vital element often came-up to maintain and to safeguard the socio-economic power. It is not irrelevant to observe here that the social hierarchy of caste as a whole had generated economic differences in the system, besides the development of other social relations among different agrarian strata in the society.

Béteille (1969), besides studying caste, also emphasises on the study of economic and political systems. A degree of autonomy exists in the inter-caste relations with regard to the above factors. Some scholars viewed that caste is changing/transforming into class in terms of nature and extent of land holding (Kumar, 1962, 1965; Mukherjee, 1957; Mencher, 1975; Miller, 1975; Kolenda, 1978; Harris, 1982; Siva Kumar and Chitra, 1979; Mukhopadhyaya, 1980; Omvedt, 1982; Subha, 1984).

VI

The agricultural labourers, based on the nature of work, are classified as farm servants and casual hired labourers (Dasgupta, 1977: 36). The casual hired labourer takes up work in farms specific for daily wages. While working like this, he has less economic security and lower social prestige to that of permanent farm servants (Béteille, 1974: 154). Kolenda (1978: 139) has pointed out that the landowners' only obligation towards the casual hired labourer is to pay wages and they still earn more per month than the farm-attached workers do. The attached workers get (consumption) loan from the landowner with whom they working and this will not be possible for hired casual labourers (Bandopadhyaya, 1979: 12). In addition to this, the permanent labourers get more workdays per season than the casual labourers who have sufficient work only during peak agricultural operations (Mencher, 1978: 351). In India, mostly, big farmers maintain attached labourers (Bardhan, 1980: 1943). Thus, the caste system not only confines to the economic relations but also has social advantages to the agricultural labourers.

Wage conditions is another important area of relations between labourers and cultivators. The wage rates vary with the sex of labour. Women labourers are usually paid a lower rate than men did (Sengupta, 1962; and Bardhan, 1980: 1947). It was also observed that the wage rates do vary with respect to the agricultural operations (Sushila Mehta, 1980: 97). Nayar (1976: 194) points out that wage rate differ from place to place and also with different workers under the same landlord. Women play a distinctive role in rural economic activities in earning a livelihood for the family. Except in the case of rich landlords and upper castes, women are usually engaged in both the production and marketing of products of agriculture and handicrafts. Women in agriculture are concentrated in activities like harvesting, weeding and transplantation and studied in changing context (Saradomani, 1982).

The land owning classes of India differ among themselves in their caste status, size, landed property and their residential status. For a majority of the people in India, economic organisation is built up in and around agriculture. One or several caste groups usually hold the dominant economic power or position through the ownership of land and control of capital (Michie, 1973: 72). In traditional agrarian structure of India land ownership was important factor in establishing dominance in all aspects of life (Srinivas, 1966: 2), because land ownership meant not only wealth and status but also power (Srinivas, 1976: 110).

Land being concentrated in the hands of a few, which again constitutes mainly of ex-intermediaries and big peasants, plentiful supply of cheap labour and stubborn persistence of earlier forms of exploitation of farm labour are bound to remain within this structure for a long period of time (Gupta, 1969: 311). Landlessness has become the root cause of exploitation of agricultural labour in India. These exploitations have taken dimensions of relations in the village, which are essentially feudalistic in nature (Konar, 1977: 56).

In the context of national economy, land relations *vis-a-vis* social structure and political systems are interrelated and internally inter-woven (Joshi, 1969: 444). The economic activities as well make them politically powerful (Prasad, 1979: 39). Ghanshyam Shah (1977: 70) found that it is only the landlords, rich and the neo-rich, who enjoy dominant position in the political and administrative processes. It was observed that the big landlords either in voting pattern or in their family lives (Shiv Kumar, 1978: 766) influence the landless labourers to a large extent. At times, it was also observed that, bondage levers these political powers (Schoer, 1977: 17).

VII

After independence government has taken the role of protection and safe guarding the agrarian masses through various development and welfare schemes which were intended to provide employment and necessities like food, clothing, etc. In the recent decades, there was a considerable change in the rural society. The various developmental schemes introduced by the government had resulted in improved educational facilities, awareness about their rights and privileges as enshrined in the constitution, through legal provisions, etc. In the context of all these developments, the social institutions got altered to a significant level and this could be perceived in different aspects of rural life.

Understanding the agrarian relations, which are dynamic, have become more complex than ever. Hence, studying the agrarian social structure in the paradigm of either inter-caste relations or class relations exclusively becomes inadequate. Some of the scholars have avoided taking either of the two extreme positions and considered caste/class model for understanding the agrarian social structure.

The introduction of a new crop in the traditional system of farming for market or export-orientation has brought about changes in the type of production to a large extent. This type of production for market or for export has altered the whole gamut of relations. It had brought about or more money into circulation, increased wages, breaking of existing relationships and establishment of new ones, and consequently changing of values in the rural areas.

The economic changes that have occurred in a multi-caste village due to introduction of market crops, like growing of vegetables for the urban market, was an important innovation. The dominant caste people despised the members of the caste who grew chestnut in tanks and ponds. The money involved made from gardening vegetables had encouraged the others to grow vegetables, stigma earlier attached for it had gradually vanished. This has led to improvement in their level of living in better quality of food, clothing and shelter. First preference given was for construction of *pucca* houses. The customary ornaments of silver were replaced by gold. Spending generously in rituals, marriages and funerals increased as these were linked to economic and social status. Gift obligations increased and children were sent to school (Rao, 1972).

The traditional way of life of Andean peasant is more intertwined with cocoa trade and the cash economic activities related to occupation from cocoa trade. The dependence of local peasant economy around cocaine brought about changes in the behaviour of the indigenous people, which was the basis of their economy. Agriculture remained totally dependent on old methods of cultivation and modes of payment, while cocoa depended on modern methods of cultivation and cash payments (Morales, 1986). The tobacco cultivation in Guntur was export oriented and the rich peasants were the ones who benefited the most. Due to increase in the area under tobacco cultivation traditional bonds were broke down, the attached labour disappeared and casual employment became the norm (Duvvury, 1986).

The introduction and expansion of commercial coconut and copra has not only changed the economic organisation from subsistence economic activity in the Tolai community of Vanamami, but has brought about political consolidation of warring hamlets and integrated them into larger villages. This also led to concentration of Wealth in a few hands. Further, dependence on skilled labour from outside has substantially increased (Salisbury, 1961).

VIII

The review brings to the fore the significance of various factors responsible for the changing agrarian relations in rural India. Further, it also points out the significance of different approaches in studying agrarian societies. In view of this the present study in a village in East Godavari district of Andhra Pradesh will provide a different dimension to the changing agrarian relations under the influence of aquaculture.

The introduction of Green revolution in the later part of 60s and cash crops during 1970s were intended to meet the demands of the ever-increasing population. They required more labour to work in various related activities. It had increased the incomes of the landowners to a large extent as well as the workdays for labourers initially. Among the landowners, Green revolution was more beneficial to the large and medium farmers and cash crops benefited the marginal and small farmers.

In the later part of 80s, pisciculture was introduced in the coastal parts of Andhra Pradesh where there is abundance of water. The indulgence with pisciculture was observed mostly with the upper caste and class farmers, as this needed large investments. This needed minimum labour as compared to agriculture and the production was to meet the demand for fish within the country.

The liberalisation and opening up of economy has led to many changes in the field of industry, trade, and investments for infrastructure had its influence on the economic structure. In this situation, aquaculture was introduced in the coastal areas and the Government encouraged this for export to earn foreign exchange. Prawn/shrimp has a greater demand in the international markets. Prawn aquaculture needs saline water from estuaries and backwaters, and coastal areas are suited for the rearing of prawn. Many coastal villages in the rural areas have converted agriculture lands to aquaculture ponds. It is only the land owning caste/classes who have adopted it as this requires huge investment for conversions of land and other accessories related to it.

Aquaculture is altogether a different type of production system. It has little semblance with agriculture. This needs new skills for rearing prawn. The landowner's who were lured by the huge profit margins converted their agricultural lands to aqua farms. As a result, the landless agricultural labourers are forced to work on these farms because of absence of employment opportunities elsewhere. The farmers or the labourers working on it did not possess any formal training or skills. Aquaculture, when compared to agriculture, supports fewer labourers who are needed in monitoring and supervision of the ponds.

The labour thus had to adjust to the changing situation. With the new forces of market economy entering into the agrarian scene, it provides an opportunity to study and understand the production pattern geared to meet the demands of international market. It is also important to understand the processes of both agriculture and aquaculture in order to analyse the change in the agrarian relations in the village under study.

Any change in the means of production would alter the existing relationships of production. It would, in turn, result in associated changes in the domain of agrarian social structure. The theoretical perspective adopted in this study regards that if there is no change in the ownership of means of production, it might not alter in any significant way the existing social structure and relations of production. The actors in the production process of agriculture are also the same as in the new production strategy of aquaculture. The changes that could be observed are only kind of adaptive changes and they can not be regarded as structural.

IX

For the selection of the study village the Mandals in the district of East Godavari having access to the backwaters were first considered, as they are important for prawn/shrimp culture. Utmost care was taken to consider the Mandals having total irrigation facilities and which are totally rural in character with backwaters. Further, the density of population, male-female ratio; literacy; the proportion of scheduled caste population to the general population; the occupational details, such as 'farmers', agricultural labour and other workers; of different Mandals were analysed to select a Mandal for the purpose of the present study.

As these Mandals were located near and along the coast, they have lands under salt pans (*parralu*[9]), wastelands and agricultural lands. Agricultural production on the above lands is low. However, the lands converted to aquaculture farms are highly productive agricultural lands and significantly they do not belong to any of the above mentioned categories. Since, our study proposes to understand the changes in agrarian relations in a predominantly agricultural village where aquaculture is significantly practised, after a close scrutiny Allavaram Mandal was selected.

Again the same attributes were taken into consideration for selection for selecting a village for the purpose of our study. Thus a multi-caste village, Godilanka, was selected. It is essential to mention here that this village does not have the fisherman.

Objectives of the Study

The present study intends to

1. Study the intra- and inter-caste relations, occupational structure, land-labour relationships, patron-client relations, etc., which are seen as a part of village social structure;
2. Understand the factors influencing the shift in production process, i.e., from agriculture to aquaculture;
3. Delineate socio-economic conditions before and after the emergence of aqua culture;
4. Assess adaptive strategies in the new milieu both by the land-owning and the agricultural labour castes; and
5. Identify the caste/class groups that have benefited from the new production strategy.

Methodology

The present study is a micro study conducted in a multi-caste village. It is conducted in the backdrop of the introduction of prawn/ shrimp aquaculture in this region, in general, and in the village under study in particular. The study adopted both qualitative and quantitative methods for the collection of data. .

Initially, a pilot survey was carried out during April 1997 to in this area. During this period, details of land such as cropping pattern, area under paddy as well as prawn/shrimp culture were collected. Demographic details pertaining to the population distribution including caste composition and occupational categories were gathered for an initial understanding of the village. This has facilitated the designing of the study tools. After this, an intensive fieldwork for a period of 12 months from June 1997 to July 1998, covering the annual cycle encompassing all seasons of agriculture and aquaculture, was carried out. Anthropological field methods and techniques were employed for the collection of data.

Census details pertaining to household, family, age-sex composition, marriage, educational qualifications, land holding pattern, occupation, income, wages, livestock, material possessions of various caste groups were collected at individual level.

Staying/living in the village has given an opportunity to observe, inquire and collect information about the various activities and their relations. This was carried out at both individual level and at group, both formal and informal levels. Initially, the villagers were reluctant to furnish information. There are important reasons for their reluctance. The cyclone of November 1996 that struck the coast of East Godavari has had an immense impact both on property and life. The experience of Godilanka was limited to only destruction of coconut gardens. The State government has sanctioned relief to the affected families. The villagers had reported that the relief money did not reach the farmers whose coconut gardens were actually affected. Later, the State Government had set up an inquiry committee and found the allegations were true. Following the recommendations of the committee, the government has ordered the villagers to pay back the relief money, so that it can be reallocated.

The timing of fieldwork thus had its initial set back. The villagers, especially the landlords were unwilling to co-operate as they had suspected the researcher to be a government agent/official. However, good rapport was established after lengthy discussions and explanations, regarding the purpose of the study. The tenants and landless agricultural labourers showed keenness in furnishing the information explaining their situation and prevailing circumstances.

It is a coincidence that as the data collection was in progress, some of the Government lands in the adjacent village were converted into aquaponds and this was published in vernacular newspapers, which drew the attention of the officials concerned. Some of the kin members of the residents of Godilanka were also involved in this type of activity. Government initiated action to protect the government lands. This also affected data collection at this juncture.

However, stay in the village developed trust and confidence among the villagers. This had facilitated the interaction and thus the flow of information became relatively easy. The villagers, both the Rajulu and others, were convinced of the intentions of the research and its importance. In fact, they have offered a rent-free accommodation and occasional lunches and dinners were extended by all the caste groups in the village. There were still some areas where the pace of information flow was slow. A few wealthy landlords felt uneasy in discussing land assets and relations with groups and lobbies within and outside their caste groups.

Participant observation was the most important method adopted in our study. It was supported by other methods such as the informal interviews, case studies, informal interviews using checklists, etc. The key informants were selected from various caste and age groups. Due importance was given to all the caste groups, which are spatially isolated in the village, for formal interviews and group discussions. The discussions and informal talks had facilitated to obtain the attitudes and views of their caste and class, especially the weaker sections which they felt, would be handicapped while talking individually or at the place of work.

Case study method was employed to understand and analyse the earlier and present situations/circumstances. Genealogies were collected pertaining to the patron-client relations or traditional relations in both the agricultural and aquaculture production systems.

Secondary information from official sources at district, Mandal, and village levels were collected. Household is taken as a unit for analysis, although due importance is given to caste groups in the village.

Chapterisation

The present study is dealt in six chapters. The first chapter throws light on introducing the reader to the topic and the area of study. It contains a brief review, objectives, methodology, and chapterisation of the thesis.

The second chapter deals with the village profile. Here social, economic, and political spheres in the study village are covered. In this chapter land, people, relationships among the different caste groups in the village, etc., as a part of social structure are discussed.

Agriculture, covering agrarian relations, and social structure are the main focus of the third chapter. The occupational categories of landowners, sharecroppers, tenants, and agricultural labourers are dealt with in detail. Due importance is given to the role of credit, which is a crucial aspect determining the patron-client relations in Godilanka, in this chapter. It also focuses on traditional payments to the service caste and their continuation. Further, it discuses the role of moneylender, such as *sahukar*, commission agent, and big farmer, in the agrarian context today in Godilanka.

The main emphasis of the fourth chapter is on aquaculture. The emergence, the development trends, and the factors that led to the growth of aquaculture in the region as well as in the village are vividly discussed in this chapter. The categories involved in the production of aquaculture, viz., owner, tenant, labourer, and their relations are dealt with in detail in this chapter. The effects of aquaculture on environment, both physical and social, are also analysed in this chapter.

The fifth chapter is on socio-economic change. This deals with the perceived changes that are observed on account of introduction of aquaculture in Godilanka. Changes in the economic organisation and its impact on the life style and standards of living, inter-caste relations, intra-caste relations, etc., are discussed in this chapter. Due importance was also given to the changes that have been taking place in the political sphere.

The study findings are briefly summarised in the concluding chapter. This chapter also provides the theoretical and other implications of the study. Besides its contribution to anthropological literature concerning agrarian relations, agrarian structure and the changes that have been taking place, the study provides suggestions and recommendations for the use of planners, administrators and development workers.

NOTES

1. *Zamindari* system.
2. Land revenue in 1793; In 1794 the system of provincial councils was replaced in the by Collectorates (Bava, 1969: 327). Permanent settlement

was introduced in this district in 1802-3 (ibid.).

3. The system of revenue collection by *ryotwari* settlement for the new Godavari district was between 1860 and 1867(Bava, 1969: 329).

4. *Jajmani* system is an economic arrangement expressed in social relationships (patron-client) in a village with its network of exchange of services and goods (Wiser, 1936; and Omvedt, 1980).

5. After his work which was carried out during 1920 along with his wife in a village referred to as Karimpur situated United Provinces of North India, now in Uttar pradesh.

6. Hilly region in the Western Ghats along the West Coast of India of Mysore state, South India.

7. Kangra is a Himalayan district in Himachal Pradesh (north of Punjab) in northern India.

8. Introduction of sericulture, which is an agro-husbandry, also brought about change in the social relations in the dry land areas, especially in Rayalaseema and many parts of the adjoining Karnataka State. It needed lesser capital for establishment and is practised all round the year. Considerable labour was also needed for sericulture. Mainly women labour is employed for this. Among the small and marginal farmers of high castes, it was taken as a household enterprise engaging their entire family members. This had augmented their family incomes and also limited these categories of farmers to be wage labourers during the lean sericulture season.

9. The lands around the shore, which receive backwaters, depend on the tidal movement in the sea (high tide). It takes a long time for this water to recede or sink and are marshy.

CHAPTER 2

Village Profile

This chapter, to begin with, provides a brief description of the natural resources, geographical boundaries, etc., of East Godavari district, in order to locate the study village in a wider context. An understanding of the village, with its historical background, composition, resources, amenities, economic and social organisation, etc., will provide a better perspective for the understanding of the study problem.

District Profile

East Godavari is a residuary part of the old Godavari district, after the separation of West Godavari in 1925. This district is closely associated with the river Godavari, which occupies a major portion of the delta area. It is situated on the north-east coast of Andhra Pradesh and is bounded on the north by Visakhapatnam district and the State of Orissa, on the east and the south by the Bay of Bengal and on the west by Khammam and West Godavari districts. The headquarters of the district is located at Kakinada.

Physical Features

East Godavari is located between the northern latitudes of 16° 30' and 18°, and the eastern longitudes of 81° 30' and 82° 36'. It occupies an area of 10,807.0 sq. km and has a population of 4,541,306, as per 1991 census.

The general elevation of the district varies from a few metres near the sea to about 300 metres in the hills of the agency. The district can be broadly divided into three zones, viz., the agency or the hill tracts, the delta, and the uplands. The Eastern Ghats rise by gradations from the level of the coast and spread throughout the agency taluks[1] of Rampachodavaram and Yellavaram. The delta portion constituting the whole of Konaseema[2] and portions of erstwhile Kakinada, Ramachandrapuram and Rajahmundry taluks presents a vast expanse of rice fields surrounded by plantain, betel, coconut gardens and innumerable palmyra.

Climate

The climate of this district is characterised by high humidity, almost throughout the year. The summer season, which starts in the middle of February and lasts till the first week of June, is harsh. This is followed by the Southwest monsoon, which lasts till the end of September. The post-monsoon or the retreating monsoon season occurs during October and November. The season is generally fine between December and mid-February.

Rainfall

The average annual rainfall of the district is 11,378 mm (44.79"). Both the Southwest and the retreating monsoons bring rains to the district. The rainfall during the Southwest monsoon, i.e., from June to September, amounts to about 61 per cent of the annual rainfall. During this period October is the rainiest month.

Flora

The vegetation of East Godavari district can be broadly divided into (a) coastal vegetation, (b) vegetation of the plains, (c) vegetation of the low lying hills, and (d) vegetation of the hills.

The mangrove forests of Koringa Reserve are situated on a flat swampy ground in the midst of salt-water creeks. In mangrove species, the seeds germinate on the tree itself, which is known as vivipary and when these germinated seeds fall on the ground they develop root system and establish mangrove forests.

The vegetation on the plains is mixed and dry, where as deciduous types of forests are found in the forest blocks. In low-lying hills scrub jungles of poor type are found. In the hills vegetation is good and miscellaneous dry deciduous and semi-evergreen forest is found in the Rampa Reserve.

Fauna

The district has representation of all classes of animals that are normally found in the forest area, though numerically less. Gaur, Sambar, Spotted Deer, Barking Deer, etc., are found in the interior forest. Occasionally, Tiger and Panther are also found. Rabbit, Hare and Wild Boar are also found in these areas. Among avifauna, Peafowl, Jungle Fowl, Partridge, Quail, and Green Pigeons are found. The mangrove forest has seagulls, cranes, herons and geese during winter.

Agriculture and Irrigation

Agriculture is the main economic activity in the district. The total area under cultivation is 5.93 lakh hectares. In this food crops are observed in 5.15 lakh hectares (86.9%) and non-food crops such as cotton, gingelly, tobacco, groundnut and coconut comprise an area of 0.78 lakh hectares (13%). Among the food crops paddy is the principal one (62%) followed by pulses (7.5%), sugarcane (2.4%), etc., (Bhatt, 1998:131). The main source of irrigation in this district is river Godavari.

Population and Sex Ratio

The density of population in East Godavari is 420 when compared to the state, which is 241 per square kilometre. It forms about 7% population of the state, which is 76% rural and 24% urban in character and has 998 females per 1000 males (Census, 1991).

Animal Husbandry

The total livestock population of East Godavari was 10.46 lakhs (3.35% of the state). Most of the cattle are indigenous and non-descript breeds. The district occupies a prominent place in the fisheries resources, both marine and inland. There are marine and estuarine fishermen villages with 104 fisherman co-operative societies of which 27 are marine fisherman societies and remaining are inland fishermen co-operatives. There are 11 fish seed experimental farms in the district.

With the advent of aquaculture along the coast of the district, from Kakinada to Tuni, about 40-45 prawn hatcheries along with their experimental farms had been established. These hatcheries supply the post-larval forms to local aqua farmers and also to neighbouring districts (even to Nellore district).

Industries

The industries in large-scale sector found are sugar mills, spinning and textile mills, electrical and paper manufacturing units. The ceramic and graphite crucible industries are in the small-scale sector. There is no heavy industry in this district.

Godilanka is located in Allavaram Mandal (previously in Amlapuram taluk) of East Godavari district. It is about 6 Kms from Amalapuram by road and 65 Kms from Rajahmundry, the nearest trade centre. It is approximately 110 Kms from the district headquarter, i.e., Kakinada. Godilanka is situated at a distance of three kilometres from the Bay of Bengal coast.

Village History

The residents of Godilanka consider that their village is of recent origin. They say that it has a history of only four to five centuries. For a better understanding of the village and its history, it is important to deal with its origins, growth, and composition.

The present village was established on the *lanka* lands of the adjacent Godi village. Godi is situated on the banks of Vasista Godavari (a tributary of Godavari) which joins the Bay of Bengal, three kilometres away from the village. The *lanka* lands are formed by the river deposits (Imperial Gazetteer, 1908:288). There is a relatively low lying area which gets water logged during rainy season due to poor drainage. *Lanka* lands will be the first to be inundated, either partially or totally, during floods. They, in general, serve as grazing grounds for the neighbouring villages.

It is believed that a few generations ago some families belonging to *Rudraraju* lineage of Kshatriya caste had migrated from Narasapuram (presently in West Godavari district) to Sakinetipalli. Competition for land and resources forced some of the *Rudrarajus* to migrate along with other lineage members. Among them, a few had settled in Godi. Two brothers from *Rudraraju* lineage had moved to the *lanka* lands east of the village from Godi (VarahalaRaju, 1980:87). They levelled the area and made it suitable for human habitation. The habitation or village was thus formed after reclamation of *lanka* lands of Godi, hence the name Godilanka.

The Kshatriyas (hereafter referred to as Rajulu, as they are locally called) of the *Rudraraju* lineage are all descendants of one of the two brothers (as one of them had no children), who are considered as founding fathers of the village Godilanka. The place where the 'founding' fathers settled is known as Pathuru. There is a big tank in Pathuru and it is the main source of water for the people settled in its vicinity. This is known as Pathuru Cheruvu. In the past Rajulu used to build houses around Pathuru Cheruvu.

Demographic Distribution

There is an anomaly with regard to the population figures of the village. According to the 1991 census, the population of Godilanka was reported as 1,802 with 424 households. A survey by the District Collector was conducted under Akshara Godavari (a literacy campaign programme) through the Department of Education. The demographic details obtained from Akshara Godavari (conducted in January 1994) indicate that Godilanka has a population of 1,631 inhabiting in 337 households. According to the census data collected by the researcher in 1997-98, the population of Godilanka was 1,450 living in 325 households. Hindus and Christians are the two religious groups found in Godilanka. Hindus (92%) predominate over Christians (8%). The Christians are converts from the Schedule Castes, especially from Mala caste.

The village has a significant proportion of Rajulu households (37.2%), the dominant caste. Except for Mala (33.5%) and Settibalija (13.8%), all the other castes are numerically insignificant, though not functionally. The castes like Golla (2.1%), Kapu (1.8%), Mangali (3.4%) and Madiga (8.0%) play a very notable role in the corporate functioning of the village (Table 2.1).

Marital Status

In Godilanka, more than half of the population among all castes is married. The unmarried constitute more than a third of the population in all caste groups. However, the proportion of divorcees, widows and widowers are marginal (Table 2.2).

Age Groups

Three-fourths (74%) of the population are below 40 years. Among different age groups, 11-20 years (20%) age group has a higher proportion as compared to 21-30 years (19%), 31-40 years (17%), 41-50 years

(11%) and 51-60 years (9%). There is a considerable proportion of members below 10 years (18%) and those above 60 years (6%) are not very conspicuous as compared to all the other age groups. These trends are more or less similar among different caste groups and also between sexes (Table 2.3.1 and 2.3.2 on page 40).

Table 2.1: Population of Godilanka

Caste	No of HHs	POPULATION		
		Female	Male	Total
Rajulu	121(37.2)	255 (49.3) (35.8)	256 (50.6) (34.6)	511(100) (35.2)
Kapu	6(1.8)	8 (42) (1.1)	11 (57.9) (1.4)	19 (100) (1.3)
Golla	7(2.1)	18 (50) (2.5)	18 (50) (2.4)	36(100) (2.4)
Settibalija	45(13.8)	93 (45) (13)	111 (54.9) (14.9)	204 (100) (14)
Mangali	11(3.4)	21 (48.8) (3.1)	24 (51.1) (1.7)	45 (100) (3.1)
Mala	109(33.5)	251 (49.2) (35.9)	263 (50.7) (34.9)	514 (100) (35.4)
Madiga	26(8)	57 (47.9) (8.2)	63 (52) (8.4)	121 (100) (8.3)
Total	325(100)	703 (48.4) (100)	747 (51.5) (100)	1450 (100) (100)

Figures in parenthesis are percentages.

Spatial Distribution

Godilanka can be broadly demarcated into three main areas: the upper, central and lower parts. The upper part of the village houses families belonging to Settibalija caste. Rajulu predominantly occupy the central part. Here, in addition to Rajulu, a couple of Mangali (Barber) and Settibalija caste families also live in the central part. The lower part is totally inhabited by Scheduled Caste families belonging to Mala and Madiga castes. The upper, central and lower parts in the village are demarcated relatively on the basis of general elevation of the areas within the village (see Map 2.3).

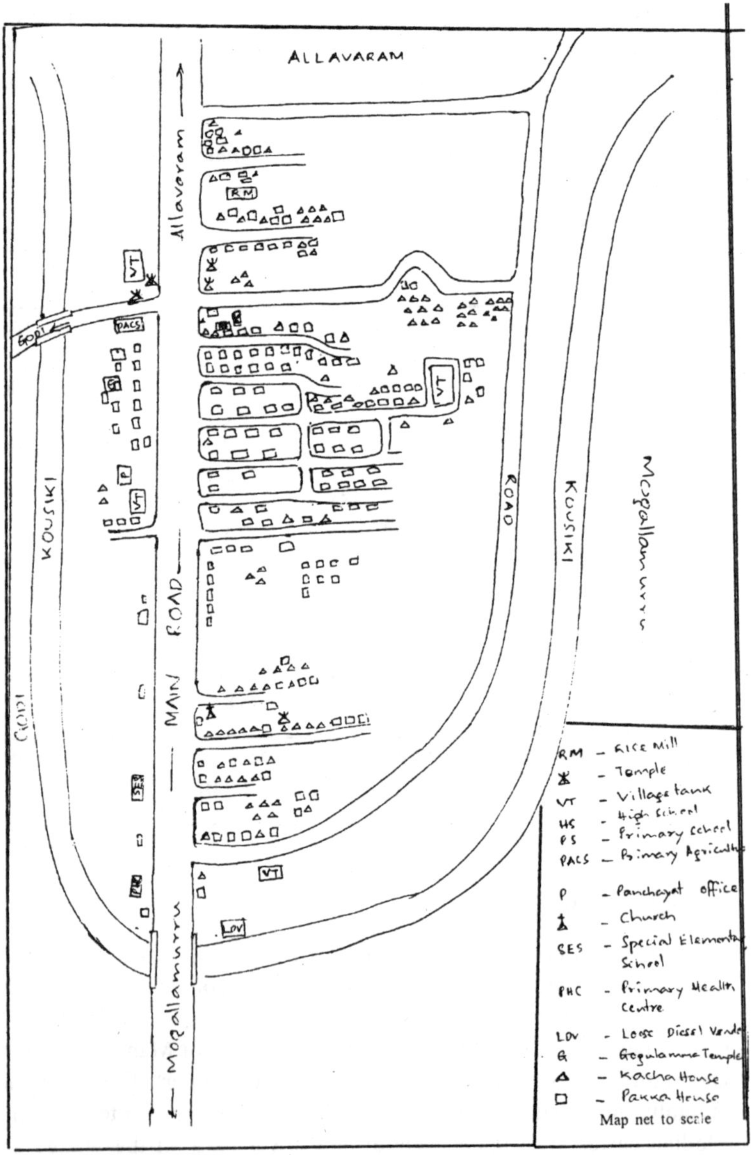

MAP 1. HOUSEHOLD PATTERN OF GODILANKA

Table 2.2: Marital Status among Different Caste Groups

Caste	Marital Status					Total
	Married	Unmarried	Widow	Widower	Divorcee	
Rajulu	334	155	18	2	1	511
	(65.5)	(30.4)	(3.5)	(0.4)	(0.2)	(100)
Kapu	11	7		1		19
	(57.9)	(36.8)		(5.3)		(100)
Golla	20	15	1			36
	(55.5)	(41.7)	(2.8)			(100)
Settibalija	120	79	3	2		204
	(58.8)	(38.7)	(1.5)	(1)		(100)
Mangali	24	18	3			45
	(53.3)	(40)	(6.7)			(100)
Mala	252	236	19	7		514
	(49)	(46)	(3.7)	(1.3)		(100)
Madiga	60	58		2	1	121
	(49.6)	(48)		(1.6)	(0.8)	(100)
Total	821	569	44	14	2	1450
	(56.6)	(39.3)	(3)	(1)	(0.1)	(100)

Figures in parentheses are percentages.

Table 2.3.1: Age-group break-up of Village Population (Males)

Age group	Rajulu	Kapu	Golla	Settibalija	Mangali	Mala	Madiga	Total
0-10	35	1	4	15	2	53	16	126
11-20	43	3	4	24	6	67	14	161
21-30	50	1	2	25	6	46	7	137
31-40	48	2	3	18	1	34	11	117
41-50	31	2	2	15	6	24	6	86
51-60	22	2	2	10	2	23	4	65
> 60	27	-	1	4	1	16	6	55
Total	256	11	18	111	24	263	64	747

Table 2.3.2: Age-group break-up of Village Population (Females)

Age group	Rajulu	Kapu	Golla	Settibalija	Mangali	Mala	Madiga	Total
0-10	45	1	5	17	2	53	15	138
11-20	30	1	2	15	6	58	12	124
21-30	54	1	3	20	2	48	8	136
31-40	48	4	4	19	4	30	12	121
41-50	29	-	1	9	3	31	5	78
51-60	30	1	1	10	3	20	3	68
> 60	19	-	2	3	1	11	2	38
Total	**255**	**8**	**18**	**93**	**21**	**251**	**57**	**703**

The Upper Part

The upper part of the village is on the northern side of the central part. Two groups of houses can be observed in the upper part. One group is located near the irrigation canal, which divides the central and upper parts of the village. These houses exclusively belong to Settibalija caste group. Among them, families belonging to *Vasarla* lineage outnumber the other lineages. Hence, it is referred to as Vasarlagudem (Gudem means place of living/dwelling). Vasarlagudem has a temple of

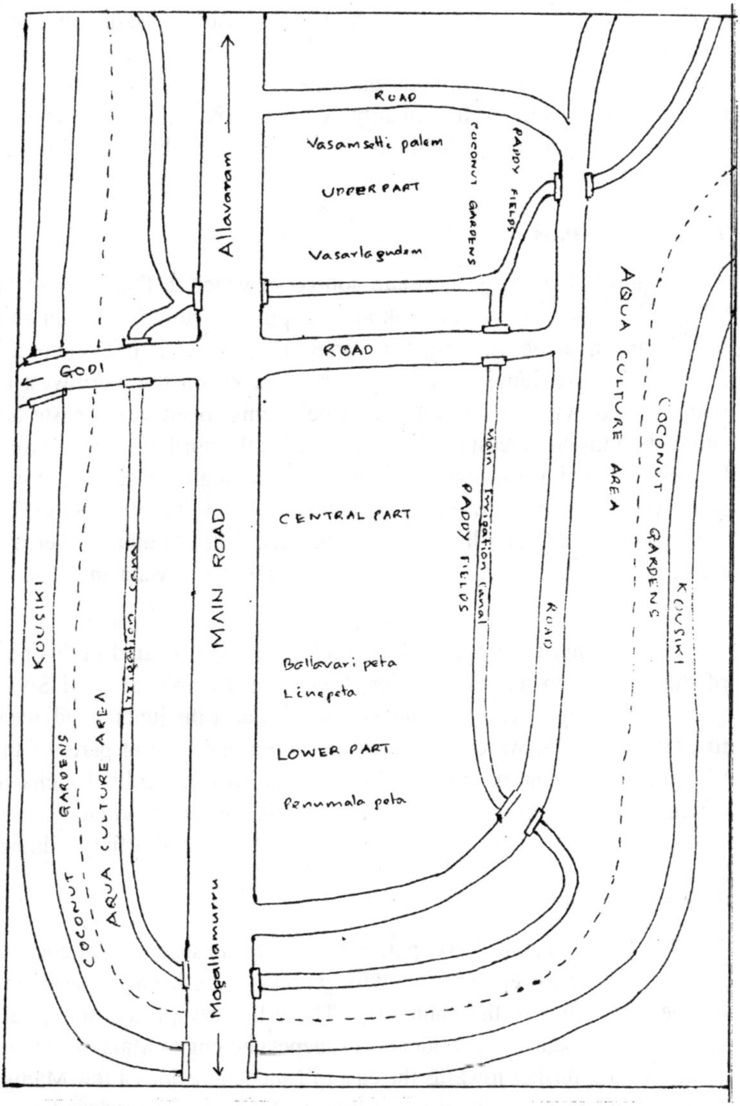

MAP 2. SPATIAL DISTRIBUTION OF GODILANKA

Goddess Kanakadurga, the presiding deity of Settibalija. To the north of Vasarlagudem and nearer to the village boundary a group of families of *Vasamsetti* lineage belonging to Settibalija caste live and, hence this area is referred to as Vasamsettipalem (Palem means settlement). In addition to them, a few Golla, Mangali, and Scheduled Caste (Madiga) families also reside in this locality. Vasamsettipalem has a rice mill owned by a Rajulu person and persons from this locality are employed in it.

The Central Part

Pathuru Cheruvu is the main source of water in this part of the village. This is a man-made tank and is approximately 5 acres in area and square in shape. An irrigation canal brings water into it. There are a few families belonging to Golla caste located on the Northwest of Pathuru Cheruvu and a few families belonging to the Kapu caste are situated on the Southeast of this Cheruvu. On the northern side to Golla houses there are a few houses belonging to Settibalija caste and on the southern side live a few Mangali families. A few families belonging to the Scheduled Castes live in the northeastern side of Pathuru Cheruvu. Here, two groups of houses can be observed wherein Mala and Madiga families reside.

Gogulamma Gudi (the village deity temple) is located in this part of the village. In addition to this, Sivalayam (temple of Lord Siva), Anjaneyaswamy *Kovela* (temple) is also seen near the junction adjacent to irrigation canal which separates the upper and central parts of the village. An elementary school, a high school, two rice mills, Panchayat office, and PACS building are located in this part of the village. There is another village tank located by the side of Panchayat office building.

The Lower Part

The central and the lower part of the village are separated by coconut gardens. In the lower part of the village, houses are linearly placed and are perpendicular to the main road. These linearly placed houses are called as 'Linepeta'. The residents of Linepeta belong to Mala, scheduled caste. A little further towards the east of Linepeta resides a few Madiga families and they are known as Linepeta Madiga. There is a line of houses parallel to Linepeta, between the central part of the village and the Linepeta, referred to as 'Ballavaripeta'. The Scheduled Caste families belonging to Mala caste from the *Balla* lineage, who are economically better off than the others, live in this line of houses. To the south of Linepeta, there is Penumalapeta wherein a group of families belonging to Mala caste from *Penumala* lineage live. The members of this locality are descendants of a single individual and are agnates. Thus, kinship played a significant role in the settlement pattern.

There is a Rama temple in Linepeta. A church is located in between Linepeta and Ballavaripeta. Besides a village tank, a special elementary school is also situated near Penumalapeta. Primary Health Centre, which provides its services to the entire Mandal, is located in the lower part of the village near the village boundary.

Social Structure of the Village

The position of the caste groups in the local hierarchy arranges the villagers into distinct status entities. While the caste groups are hierarchically graded, they are integrated into an organic whole through some economic and socio-cultural mechanisms. The various mechanisms of interdependence, which lead to the integration of caste groups, also result in the unity and solidarity of the village.

Castes are hereditary, endogamous and ranked groups. Each caste is traditionally associated with a specific occupation. However, secondary occupations are also practised. The local term for caste is *'kulam'*. The villagers demarcate their caste people as '*ma vadu* (our man) or '*ma vallu* (our people). The different castes found in Godilanka are: 1) Rajulu, 2) Kapu, 3) Golla, 4) Settibalija, 5) Mangali, 6) Mala and 7) Madiga.

Rajulu, who belong to the Kshatriya order in the *Varna* hierarchy, occupy the top position in the local caste hierarchy. They own most of the arable land in the village. Majority of them do not work on land but lease-out either to intermediaries or directly to farmers for rent. In cases where they are engaged in farming, though not directly working on the land, they employ hired labour and practice agriculture.

The Kapu comes next to Rajulu caste in the local hierarchy. They are an agricultural caste traditionally. In this village they leased-in agricultural lands and practice agriculture. A few of them also work as hired wage labour. Women from this caste group work as maidservants in Rajulu households.

The Golla (shepherd caste) ranks next to Kapu in the village. Traditionally, they were a cattle-rearing community (sheep, goat, cows,

The Golla (shepherd caste) ranks next to Kapu in the village. Traditionally, they were a cattle-rearing community (sheep, goat, cows, etc.). Initially, they were working for Rajulu on an annual payment basis and were looking after the cattle, milching cows and buffaloes, draw or fetch drinking water, assist in domestic work; and also help during rituals and festivals in the houses of Rajulu. Presently, they work as farmers on their own lands, involve in agriculture labour for wages, petty coconut business, etc. It is essential to note that women do not go for work outside their homes.

Next to Golla in the local caste ranking comes Settibalija. Their traditional occupation was toddy tapping. Previously, in Godilanka they were engaged in toddy tapping and worked as farm supervisors/farm labourers. Besides toddy tapping, they were also employed in coconut harvest, as they were experts in climbing trees. Later they shifted their occupation from coconut harvest to agriculture. They felt that climbing trees always involved risk and so availability of land gave them an opportunity to go for agriculture. Today some of them are self-employed, work as cultivators (own or leased-in), wage labourers in rice mills and agricultural fields in the village. Women work as labourers in the agricultural fields. The economic position of the particular family determines whether women of their families should work in their family farms or as labourers in other's fields.

Mangali, the traditional barber caste, comes next to Settibalija in the order of local caste ranking. About two decades ago, some younger members belonging to this caste viewed their *kulavrutti* (caste occupation) as demeaning, and switched over to tailoring. Thus, today quite a few of them are tailors. However, due to demand for their traditional occupation some younger members from this caste moved back to their *kulavrutti*. There are some from this community who give allopathic medicines and injections to the villagers, and are addressed by the villagers as 'Doctors'. They were earlier working under some qualified doctors as helpers or Compounders. However, they do not possess any formal qualifications. Interestingly, each engaged in the above occupations has his specified areas of operation within the village on caste and gender lines. Women from this caste do not work outside their homes. Occasionally, some of them work on festive occasions or during life-cycle crises in Rajulu houses.

In the lowest rung of the caste ranking in the village are the Scheduled Castes. The Scheduled Caste population comprises of Mala and Madiga

Table 2.4: Family types in Godilanka

Family	Rajulu	Kapu	Golla	Settibalija	Mangali	Mala	Madiga	Total
Nuclear	31	4	3	33	8	73	18	170
	(25.6)	(66.6)	(42.8)	(73.3)	(72.7)	(66.9)	(69.2)	(52.3)
Supplementary	24	1	2	5	3	27	7	69
Nuclear	(19.8)	(16.6)	(28.6)	(11.1)	(27.3)	(24.7)	(26.9)	(21.2)
Lineal joint	44	1	-	4	-	6	1	56
	(36.6)	(16.6)	-	(8.9)	-	(5.5)	(3.8)	(17.2)
Collateral Joint	6	-	2	2	-	3	-	13
	(5)	-	(28.6)	(4.4)	-	(2.7)	-	(4)
Lineal Collateral	16	-	-	1	-	-	-	17
joint	(13.2)	-	-	(2.2)	-	-	-	(5.2)
Total	121	6	7	45	11	109	26	325
	(100)	(100)	(100)	(100)	(100)	(100)	(100)	(100)

Figures in parentheses are percentages.

castes. Traditionally, Mala were engaged as agricultural labourers, village servants and watchmen. They are still engaged as hired labourers or as tied labour by the landowners. However, some of them are in services like teachers, nurses, postmaster, mechanics, drivers, etc. There are also some among them who practice agriculture on their own lands. The Madiga are traditionally leather workers and were engaged in cleaning carcasses and leather tanning. At present many have left their traditional occupation, as it was regarded as a demeaning occupation, and are now engaged in coconut harvesting. A few among them have also taken to toddy tapping, agriculture, cattle business, etc. Mala consider Madiga to be inferior in position because of their traditional occupation. Women from both Mala and Madiga castes work as daily wage labourers and maidservants. Women belonging to better off sections among these castes do not work outside their homes.

Family Type

To analyse the type of family Kolenda's (1968) typology was adopted. However, only a few types of families from her typology are observed in this village. More than half of the families in the village is nuclear (52%). Among the rest, supplemented nuclear (21%) and lineal joint (17%) families predominate over lineal collateral (5%) and collateral joint (4%) families.

In all the caste groups in the village, except Rajulu, majority of the families are nuclear followed by supplemented nuclear families. Among Rajulu, more than a third of the families are lineal joint families (36%) followed by nuclear families (26%), supplemented nuclear families (20%), lineal collateral joint families (13%), and collateral joint families (5%). Thus, it is significant to note that more than a half of the families among Rajulu belong to different types of joint families (Table 2.4). Among Rajulu, especially the landowners, the married sons live with their parents in the same house, in different rooms, till the property is divided. The property usually gets divided after the youngest son is married.

Size of the Family

The average size of the family in Godilanka is 4.5. The highest average size of the family is seen among Golla (5.1) and the lowest is observed among Kapu (3.2) castes (Table 2.5). Though the lineal joint families among Rajulu are more, the average size of the family is lower than the average size of the village. Although this appears anomalous, this is because the brothers live in different portions with in the same house, as they have separate hearths and, more or less, live in nuclear settings. Their property is still undivided and they consider themselves to be *ummadi* (joint) sentimentally. In these types of households either father or the eldest son are the key persons in decision making process.

Table 2.5: Caste-wise Prevalence of Size of the Family

Caste	Houses	Population	Average family size
Rajulu	121	511	4.22
Kapu	6	19	3.16
Golla	7	36	5.14
Settibalija	45	204	4.53
Mangali	11	45	4.09
Mala	109	514	4.71
Madiga	26	121	4.65
Total	325	1450	4.46

Marriage

The most common type of marriage observed is arranged marriage by mutual consent. Monogamy is the predominant form of marriage observed in the village. In Godilanka, once a girl is nubile the parents of the groom-to-be approach the girl's party and initiate the process of negotiation. The age at marriage is not same for all caste categories and it varies from caste to caste. Even within a caste, it varies depending on the economic position and the educational status. That is, the higher the class status the higher is the age at marriage.

Once both the parties reach to an understanding the alliance would be finalised. The marriage will take place according to the custom prevalent in a caste. Village exogamy, by and large, is the norm. Among all the castes, non-preferential marriages (locally termed as '*naturikam*') are more prevalent than preferential marriages. Preferential marriages, though less in proportion, are observed among cross-cousins (*menarikam*) and between mother's brother and elder sister's daughter (Table 2.6). Dowry is observed as a part of marriage alliance. This is noticed irrespective of caste and class backgrounds. Divorce is not of common occurrence and in the entire village only two cases were observed, one each among Rajulu and Madiga caste. Widower is usually encouraged to re-marry. However, the same attitude is not observed in case of widows.

Table 2.6: Types of Marriages

Caste	MBD	FZD	ZD	Relatives	Not Relatives	Total
Rajulu	13	2	3	8	141	167
	(7.8)	(1.2)	(1.8)	(4.8)	(84.4)	(100)
Kapu	-	-	-	-	6	6
					(100)	(100)
Golla	-	-	-	1	9	10
				(10.0)	(90)	(100)
Settibalija	2 3	-	2	4	52	60
	(4)		(3.4)	(6.7)	(86.5)	(100)
Mangali	1	-	-	2	9	12
	(8.3)			(16.6)	(75.1)	(100)
Mala	12	-	2	3	102	126
	(9.5)		(1.6)	(2.3)	(81)	(100)
Madiga	2	7	-	1	26	30
	(6.6)	(5.6)		(3.3)	(86.8)	(100)
Total	**29**	**10**	**7**	**19**	**345**	**411**
	(7.0)	**(2.4)**	**(1.7)**	**(4.6)**	**(84.3)**	**(100)**

Figures in parentheses are percentages.

Ownership of House

The ownership and the locality of a house are an important aspect in the village. To a large extent, one's social position in the village actually hinges on it and it further reflects one's economic status as well. Those who do not own any house, especially among the Scheduled Castes, will be at the mercy of the landowners in whose lands they are permitted to live and are obligated to work for the landowners on unfavourable terms.

A semi-*pucca* house is a category in between *kutcha* and *pucca* houses. The walls are built with cement and brick and the roof covered with palmyrah leaves or *jammi* grass. The floor may be either mud plastered with dung or with *napa bandalu.*

Kutcha houses usually have mud walls. In some cases, the walls are made of coconut leaves tied to bamboo poles (this type of wall is called *dhadi*, which means fence) and the roof will be covered with palmyra leaves over bamboo rafters. The roof is trapezium in shape slanting on all sides. In this type of house, the floor is mud plastered with dung. This type of *kutcha* house is called *gudise.*

Table 2.7: Caste-wise Ownership of Houses

Caste	Houses		Total
	Not Owned	Owned	
Rajulu	10 (8.2)	111 (91.8)	121 (100)
Kapu	1 (16.6)	5 (83.4)	6 (100)
Golla	-	7 (100)	7 (100)
Settibalija	6 (13.3)	39 (86.7)	45 (100)
Mangali	1 (9)	10 (91)	11 (100)
Mala	11 (10)	98 (90)	109 (100)
Madiga	10 (38.4)	16 (61.6)	26 (100)
Total	**39 (12)**	**286 (88)**	**325 (100)**

Figures in parentheses are percentages.

Rajulu are landowners and majority of them possesses houses of their own. Among them who do not have a house live in their relatives' houses. For instance, *Rudraraju* Krishnam Raju is living in a coconut garden belonging to his relative, with whom he is working as a supervisory farm worker.

All the Golla families are living in their own houses. The Rajulu landlords gifted the house sites to Gollas who rendered their services over generations to them. Among the Kapus, almost all, except one who lives in the land of a Raju, possess own houses. In case of Settibalija, one out of every seventh family does not possess a house of their own. Most of them live in their relatives' houses, while a couple of them are living in the coconut gardens of Rajulu working as watchmen of the garden and the service is negotiated for the rent of the house site. Among Mangali, except a single family, all have their own houses.

Among the Mala, one out of every ten families does not possess a house of their own. Those living in Pathuru, majority live in the coconut gardens of Rajulu with whom they are working. In all the other areas, the Mala families possess their own houses. A similar pattern is also observed with the Madiga families. In Pathuru, while a majority of them live in the coconut gardens of Rajulu, a few live in 'Ummadi Bhoomi' (joint property) of Rajulu, which is not claimed by anyone. In Linepeta, Penumalapeta and Ballavaripeta Malas live in their own houses (Table 2.7).

House Type

House type is classified based on the kind of material used for the construction of walls and roofs. The houses in the village can be broadly categorised into (1) *Pucca*, (2) Semi-*Pucca* and (3) *Kutcha.*

The walls of a *pucca* house are made of bricks and either with cement or '*sunnam'* (lime mortar). The roof is either tiled or RCC. Traditionally *pucca* houses are made of teak with tiled roof and these houses are called '*logillu*'. The ones with iron grills and RCC roof are called '*daba illu*'. The floor is laid with unpolished stone slabs (*napa bandalu*) in case of *logillu* and polished stone slabs in case of *daba illu* (Plate 2.9).

There exists a correlation between the type of house and the socio-economic status of a family. The higher the socio-economic position the greater is the chance of having *pucca* houses and *vice versa.*

Table 2.8: Caste-wise distribution of House type

Caste	House Type			Total
	Kutcha	Pucca	Semi-Pucca	
Rajulu	14 (11.5)	103 (85.1)	4 (3.3)	121 (100)
Kapu	5 (83.3)	1 (16.7)	-	6 (100)
Golla	3 (42.8)	3 (42.8)	1 (14.2)	7 (100)
Settibalija	22 (48.8)	22 (48.8)	1 (2.2)	45 (100)
Mangali	2 (18.1)	9 (81.9)	-	11 (100)
Mala	66 (60.5)	27 (24.7)	16 (14.6)	109 (100)
Madiga	24 (92.3)	2 (7.7)	-	26 (100)
Total	**136 (41.8)**	**167 (51.4)**	**22 (6.8)**	**325 (100)**

Figures in parentheses are percentages.

Majority among Rajulu resides in *pucca* houses. Among both Golla and Settibalija an equal proportion of families live in *Kutcha* and *pucca* houses. Most of the Kapu, Madiga and Mala families live in *Kutcha* houses. Substantial proportions of Mala families also reside in *pucca* and semi-*Pucca* houses (Table 2.8).

Number of Rooms

In addition to the ownership of house and the type of house, number of rooms available in a house reflects the status and living standard of the family. More than a third of families in Godilanka live in two-room houses (41%), followed by three-room houses (19%), five-room houses (18%) and four-room houses (15%). However, a small proportion of families still continues to live in single room houses (7%).

Majority among Rajulu resides in houses that have more than three rooms. Among the rest most of them live in two roomed houses. However, among the rest, a notable proportion of families belonging to Mala caste live in houses which have more than three rooms (Table 2.8).

Table 2.9: No. of Rooms among various Caste Groups in the village

No. of Rooms	Rajulu	Kapu	Golla	Settibalija	Mangali	Mala	Madiga	Total
1	3 (2.5)	1 (16.6)	-	1 (2.2)	1 (9)	11 (10)	4 (15.3)	21 21(6.5)
2	19 (15.7)	4 (66.7)	6 (85.6)	33 (73.3)	7 (63.6)	49 (45)	17 (65.3)	135 (41.5)
3	28 (23.1)	- -	- -	4 (8.9)	2 (18.1)	24 (22)	4 (15.3)	62 (19)
4	27 (22.3)	1 (16.6)	- -	6 (13.3)	1 (9)	13 (11.9)	1 (3.8)	49 (15)
5 & above	44 (36.3)	- -	1 (14.3)	1 (2.2)	- -	12 (11)	- -	58 (17.8)
Total	**121 (100)**	**6 (100)**	**7 (100)**	**45 (100)**	**11 (100)**	**109 (100)**	**26 (100)**	**325 (100)**

Figures in parentheses are percentages.

Having a toilet, apart from number of rooms, in the present day context of Godilanka is an indication of the status that a family enjoys. Most of the villagers use open places alongside the canals for defecation.

Most of the houses that have toilets belong to Rajulu (87%) caste. Rest of the houses, which possessed toilet had them due to a Government sponsored scheme under the Janma Bhoomi programme (Table 2.10).

Table 2.10: Caste-wise toilet facilities

Caste	Place of Defecation		Total
	Communal Open	Own Toilets	
Rajulu	41 (33.8)	80 (66.2)	121 (100)
Kapu	5 (83.3)	1 (16.7)	6 (100)
Golla	5 (71.4)	2 (28.6)	7 (100)
Settibalija	42 (93.3)	3 (6.7)	45 (100)
Mangali	10 (90.9)	1 (9.1)	11 (100)
Mala	104 (95.4)	5 (4.6)	109 (100)
Madiga	26 (100)	-	26 (100)
Total	**233 (71.6)**	**92 (8.4)**	**325 (100)**

Figures in parentheses are percentages.

Fuel for Cooking

In Godilanka, dung cakes and residues of coconut trees are used as fuel for cooking and other purposes by most of the families. Looked at caste-wise, majority of Rajulu households possesses LPG, besides using conventional fuel (Table 2.11on page 53). In Godilanka, the emergence of aquaculture coincided with increased use of LPG.

Amenities

The amenities available in Godilanka are drinking water, electricity, communication and transport.

Drinking Water

In Godilanka, the main source of drinking water is wells. Among Rajulu, most (70%) of the households possess their own wells and the

Table 2.11: Caste-wise distribution of Cooking Fuel

Caste	Fuel for cooking			Total
	Dung cakes & coconut residues	Kerosene	LPG	
Rajulu	51 (42.1)	-	70 (57.9)	121 (100)
Kapu	5 (83.3)	-	1 (16.7)	6 (100)
Golla	7 (100)	-	-	7 (100)
Settibalija	44 (97.7)	-	1 (0.3)	45 (100)
Mangali	8 (72.7)	2 (18.1)	1 (9)	11 (100)
Mala	109 (100)	-	-	109 (100)
Madiga	26 (100)	-	-	26 (100)
Total	**250 (76.9)**	**2 (0.6)**	**73 (22.4)**	**325 (100)**

Figures in parentheses are percentages.

remaining share water with their neighbours. One household each from Golla, Settibalija and Mangali and two Mala families have their own drinking water wells (Table 2.12).

Table 2.12: Drinking Water Facility of Different Caste Members

Source	Rajulu	Kapu	Golla	Settibalija	Mangali	Mala	Madiga	Total
Own well	87 (95)	-	1 (1)	1 (1)	1 (1)	2 (2)	-	92 (100)
Shared/ Public well	32 (13.8)	6 (2.6)	6 (2.6)	44 (19)	10 (4.3)	107 (46.3)	23 (11.2)	231 (100)
Under-ground bore	2 (100)	-	-	-	-	-	-	2 (100)
Total	**121 (37.2)**	**6 (1.8)**	**7 (2.2)**	**45 (13.8)**	**11 (3.4)**	**109 (33.5)**	**26 (8)**	**325 (100)**

Figures in parenthesis are percentages.

In Vasarlagudem and Vasamsettipalem, inhabited by Settibalija, drinking water is drawn from four common wells constructed under JRY programme (a rural development programme). They are located near the main road, by the side of irrigation canal. There is a well in the PACS compound and Gollas depend on this well for their drinking water needs. Kapu and Mangali depend on the neighbouring Rajulu households for their drinking water needs.

The Pathuru Mala households have a well nearer to their dwellings from which they fetch water. This well was dug with the money collected by their caste *Sangham* (council). Madiga of Pathuru have a well exclusively for themselves, which was donated by Rajulu. The Scheduled Caste population of Ballavaripeta and Linepet depend entirely on a well owned by a Rajulu household (teacher), who allows them to fetch water from his well. This well is located at about half a kilometre from their homes. The residents of Penumalapeta bring water from the wells of neighbouring village Mogallamuru, which is at half a kilometre distance from Godilanka. It is the women's job to draw and carry water from the wells to their homes.

Electricity

Godilanka is an electrified village. Almost all houses among Rajulu have electricity. Majority among all the castes, other than Rajulu, does not have electricity. Among those who have electricity, other than Rajulu, Settibalija and Mala are better placed. All the Madiga houses are devoid of electricity as their houses are located in the midst of coconut gardens (Table 2.13).

Table 2.13: Caste-wise Source of Lighting

Caste	Source of Lighting		Total
	Electricity	Kerosene	
Rajulu	119 (98.3)	2 (1.7)	121 (100)
Kapu	1 (6.4)	5 (83.3)	6 (100)
Golla	1 (14.2)	6 (85.8)	7 (100)
Settibalija	10 (22.2)	35 (77.8)	45 (100)
Mangali	3 (27.2)	8 (72.8)	11 (100)
Mala	27 (24.7)	82 (75.3)	109 (100)
Madiga	-	26 (100)	26 (100)
Total	**161 (49.5)**	**164 (50.5)**	**325 (100)**

Figures in parentheses are percentages.

Communication and Transport

There is a post office manned by a resident of Linepeta from Mala caste. A considerable number of big landlords among Rajulu have telephones at their residence. There is a public telephone facility maintained at *baddi* (petty shop) near *Gogulamma* temple.

Godilanka is on Amalapuram - Komaragiripatnam main road and the State Road Transport Corporation operates buses daily at an average frequency of two hours. Besides, there are private vans to transport people to nearby places/towns.

Infrastructure Facilities

In Godilanka the infrastructure facilities include school, primary health centre, primary agricultural cooperative society, temples and church. However, Santa is available in the adjacent village.

Schools

Godilanka has three schools. Of them, one is a high school and the remaining two are elementary schools. One of the elementary schools is a special elementary school to cater to the Schedule Caste children. The high school and the elementary school, which is adjacent to high school, are located in the central part of the village, while the special elementary school is located in the lower part of the village near Penumalapeta.

A Rajulu landlord donated land and building for the high school and it was named after him. This is a non- Plan school and has five (all male) trained teachers. It has four classrooms and some classes are taken even on the veranda. The high school does not have drinking water facility. Boys and girls drink water from a near by well. There are no toilet facilities. All the Scheduled Caste students attending the high school receive scholarships from the Social Welfare Department, Government of Andhra Pradesh. The Government of Andhra Pradesh supplies textbooks free of cost to all the students, irrespective of socio-economic background. The high school attracts students from the neighbouring villages also. Schedule Caste children are more preponderant than those other children attending the high school from Godilanka.

The elementary school adjoining the high school has a single hall and, as in the case of high school, classes are also held on the veranda.

The students attending this school are not only from this village but also from the nearby villages. Among the students attending this school, Scheduled Caste students are more in number. Most of the Pathuru Scheduled Caste students attend this school. In both the elementary schools, every student, irrespective of the socio-economic background, who has more than 80 per cent attendance, is given three kilograms of rice per month, under a scheme sponsored by the Government of Andhra Pradesh. In addition to free books provided to all students of all classes, scholarship is also given to the students belonging to Scheduled Caste.

Primary Health Centre

The Primary Health Centre (PHC) of Allavaram Mandal is situated in Godilanka. This is located in the lower part of the village, a little further to Penumalapeta at the village boundary. A landlord donated land and building for the PHC. The PHC has a doctor and 25 supporting staff members. Most of the supporting staff members are multipurpose health workers, both male and female (16 and 9). The main function of the staff of the PHC is to motivate people to undergo family planning operations and also conduct immunisation camps for the children in different villages on an appointed date every month. In addition to this, the health workers visit houses, conduct health meetings, and give medicines to the needy, in general, and to the weaker sections of society, in particular.

Very few in Godilanka utilise the services of the PHC. This village has a RMP and people prefer him to the PHC doctor. Most of the services of the PHC are utilised by neighbouring villages. The residents of Godilanka complain of non-availability of medicines in the PHC and say that attention is paid to only a few influential people in the village.

Temples

Most of the villagers observe festivals like Sankranti, Dussera, Sri Rama Navami, Ugadi, Vinayakachaviti, Deepavali, etc. There are a few temples in different locations in the village, where they offer their prayers during festival days. A Sivalayam is seen in the central part of the village. A landlord from Rajulu caste got this temple constructed in 1990. In addition to this, he had donated 0.5 acre of land to the temple for its maintenance. The priest, who hails from a neighbouring village, comes on alternate days and performs rituals in the temple. The offerings made by the devotees go to the priest and he receives payment in kind for his services from the produce of the temple land. Adjacent to the Sivalayam, there is an Anjaneya temple. The rice millers association of Godilanka had built this temple. Every Tuesday the villagers who have faith in the deity perform *bhajans* and *pooja*.

Gogulamma is a village deity and the temple dedicated to her is located in the central part of the village, opposite to the high school and nearer to the PACS building. An annual fair takes place at the temple on the following Thursday of *Ugadi*, the Telugu New-Year day. According to a legend, Goddess Lakshmi appeared in dream to Gogulraju, the son of the founder of the village, and asked him to build a temple for her. Since Gogulraju built the temple, it came to be known as *Gogulamma* temple. Earlier this temple was located near Gogulraju's house and later it was shifted to the temple constructed near the high school. Since inception the priest of this temple is a Kummari (potter) from Godi village. The descendants of each of the four sons of the founder of Godilanka from the *Rudraraju* lineage perform rituals for four days, respectively, before the annual fair. The descendants of each of the four brothers consider themselves as belonging to four '*kutumbams*' (families). On all these four days, the Kummari priest takes *Gogulamma* in a procession. In this procession, Mangali of the village beat drums. During the procession, the priest stops at each house and collects grain and/money. He takes this for his services to the temple. In addition to this, village elders pay him for his services at the annual fair, where he dances. The village elders also pay the Mangali, who beat drums while the priest dances.

A temple of Goddess Kanakadurga can be noticed in the upper part of the village at Vasarlagudem. The Settibalija had collected money from their caste men and also from the Rajulu landlords and constructed this temple in 1994. An old man from Settibalija caste performs *pooja* every Thursday. Dussera (which comes in the month of October) is celebrated with grandeur and gaiety at this temple.

In the lower part of the village in Linepet there is a Rama temple. The Malas, who are residents of Linepet and Ballavaripeta, visit this temple. The priest of this temple is from Mala caste from Linepeta. The temple will be kept opened only on Sri Rama Navami day (usually in the month of March) and rituals are performed. The residents of Linepet and Ballavaripeta contribute money for performing this ritual.

Church

A church is located in Ballavaripeta, in the lower part of the village and the residents of Ballavaripeta and Linepet attend this. The church is funded by a Missionary organisation from Kakinada. The priest is not a native of Godilanka, but resides in the church. Every Sunday, religious service takes place at this church. A few residents of Penumalapeta visit a church in the neighbouring village.

Primary Agricultural Co-operative Society (PACS)

The Godilanka PACS was registered on April 23, 1901 and started functioning from October 10, 1901. This is affiliated to the District Cooperative Central Bank, Kakinada, and the Andhra Pradesh State Cooperative Union, Hyderabad. The membership of Godilanka PACS in 1996-97 was 3,785 with a share capital of Rs. 8,74,746. The members also belong to the neighbouring villages. Among the members, 1,101 were from the upper castes, 958 BCs, 956 were SCs, and five STs. About 35% of members are women.

The PACS has an elected body with a President and nine members and it stays in office for a term of five years. The branch runs under the supervision of the Secretary, an employee appointed by the District Central Co-operative Bank. He, along with his staff members, is responsible for maintenance of accounts and records. The main function of the PACS is to provide loans to its members. It advances two types of loans: crop loans to the farmers and consumption loans to the non-agricultural members like agricultural labour, artisan, and petty shopkeepers. Members can also deposit money in either fixed or recurring deposits.

Santa (Weekly Market)

Godilanka does not have a *santa* (a weekly market) but the villagers attend the *santa* in the adjacent village, Mogallamuru, which is held on every Thursday. Vegetables and essential commodities like grains, pulses, jaggery, sugar, tamarind, onion, etc., are available. Traders bring goods from different places and sell them in the *santas* held at different important villages of the Mandal on appointed days of a week. *Santa* is held on either side of the road and every trader has an allocated space and one has to pay tax to the agent appointed by the Gram Panchayat. The Panchayat annually auctions the right to collect *santa* taxes to private contractors from the traders. Every Thursday, all villagers (farmers, agricultural labour, etc.,) make themselves convenient to procure their weekly requirements at the *santa*.

Caste and Landholding

There is a closer association between caste status and land ownership. The higher the caste status the greater is the chance of owning arable land and the lower the caste status the higher the chances of it being in agricultural labour. A similar relation between land, caste and labour was observed (Kumar, 1965; Mohammad, 1988; and Mathur, 1958). Godilanka is no exception to this. Social organisation of agriculture in Godilanka can be categorised into owners, tenants and agricultural labour. Rajulu possess most of the agricultural land (90.8%). Other than Rajulu, only Settibalija (6.2%) owns somewhat considerable amount of agricultural land (Table 2.14). Most of the tenants are from Settibalija, Kapu and Mala castes, while most of the agricultural labour is from Mala and Madiga castes.

Table 2.14: Caste-wise Distribution of Agricultural Lands (in Acres)

Caste	Agricultural Land (Own)	Aqua Land	Total
Rajulu	600 (90.8)	154.72 (93.6)	754.72
Kapu	-	-	-
Golla	4 (0.6)	1.2* (0.7)	5.2
Settibalija	40.7 (6.2)	4.7* (2.8)	45.4
Mangali	-	-	-
Mala	16.1 (2.4)	4.7* (2.8)	20.8
Madiga	-	-	-
Total	660.8 (80.0) (100)	165.32 (20.0) (100)	826.12 (100)

Figures in parentheses are percentages.
*These are situated in the neighbouring villages.

The villagers depending on the land owned by a farmer denote them into different categories, which they use in their day-to-day conversations. They are *Pedda ryutu*, who possess about 10 acres of agricultural land and a considerable quantity of coconut garden (more than 20 acres together), *Chinna ryutu*, who possess one to three acres of paddy land and one to two acres of coconut gardens (nearly less than 5 acres jointly). In between the above two categories of ryutulu are the landowners who possess land between 5-20 acres either agricultural land or coconut gardens jointly. They are connoted as *madyarakam* ryutulu. For a better understanding, these categories of landowners are differentiated into marginal, small, medium, large and big farmers/ landowners. Marginal and small landowners fall in the category of *Chinna* ryutulu, whereas large and big farmers come under the category of *Pedda* ryutulu. More than half of the landowners are *chinna* ryutulu and roughly one-tenth landowners are *pedda* ryutulu and the rest are *madyarakam* ryutulu. Except a Golla and two Settibalija landowners all the landowners, other than Rajulu, are *chinna* ryutulu (Table 2.15).

Table 2.15: Caste and Class of Farmers

Caste	Pedda ryutu	Madyaryutu	Chinna ryutu	Total
Rajulu	12 (100)	42 (93.4)	34 (65.3)	88 (80.7)
Kapu	-	-	-	-
Golla	-	1 (2.2)	1 (1.9)	2 (1.8)
Settibalija	-	2 (4.4)	8 (15.4)	10 (9.2)
Mangali	-	-	-	-
Mala	-	-	7 (13.5)	7 (6.4)
Madiga	-	-	2 (3.8)	2 (1.8)
Total	**12 (100)**	**45 (100)**	**52 (100)**	**109 (100)**

Figures in parentheses are percentages.

Economic Scenario

The mainstay of the village economy is agriculture, coconut gardens and aquaculture. The land is fertile and is irrigated through irrigation canals. Annually two crops of paddy are raised during 'khariff', which starts from June-November, and 'rabi', from December-April. In addition to this, a third crop of either pulses or fodder crop for cattle, '*rotta*', is grown by some during April-May.

Though Rajulu possess most of the arable land they do not work on it. They either lease-out or employ hired labour in raising crops. The production is not subsistence oriented. Required quantity in the produce is set apart for own consumption and the rest is marketed.

The extent under coconut garden is more than the area under paddy cultivation. Nearly almost all the coconut gardens belong to Rajulu. Coconut crop is treated as dry crop by revenue authorities. Large tracts of agricultural land have been converted into coconut gardens many years ago. Paddy fields were elevated by adding layers of soil and were converted into coconut gardens and they are about a foot higher in elevation than the paddy fields. They are seen all along the bunds of *Kousiki*[3], which is a drainage canal wherein all the water from agricultural lands drain into it. This opens into the sea (Bay of Bengal), which is three kilometres away from the village. The backwaters of the sea into *Kousiki* helped in establishing aqua farms.

The important feature of the coconut gardens is that they yield crop every month and is a source of continuous income. The investment on their maintenance is low, though the initial costs in establishing the gardens may be high, and the produce is almost consistent. It involves low labour input when compared with agriculture, where the rates of investment and profit are lower with a higher level of labour and supervision.

Aquaculture was introduced in Godilanka in 1993. Many of the agricultural fields near *Kousiki* have been converted into aquaculture ponds. The paddy fields adjacent to coconut gardens are converted into aquaculture ponds/farms. The early adopters were large progressive landowners, who belonged to Rajulu caste. The extent of aquaculture practised by the other caste groups is in leased-in ponds located outside the village.

Social organisation of aquaculture, like agriculture, includes the owners, tenants and labour. There is a marked heterogeneity in all the categories involved in prawn production, in terms of ownership, leasing pattern and the labour process (This is elaborately discussed in the chapter on aquaculture). It is important to note that, as in the case of agriculture and coconut gardens, aqua farms are mostly owned by Rajulu.

Table 2.16.1: Literacy Status of Various Caste Members (Male)

Literacy	Rajulu	Kapu	Golla	Settibalija	Mangali	Mala	Madiga	Total
Illiterate	6	1	3	9	2	34	13	68
	(2.3)	(9.1)	(16.6)	(8.0)	(8.7)	(13.0)	(20.6)	(9.1)
Primary	89	4	2	37	4	112	36	284
	(34.3)	(36.4)	(11.1)	(33.0)	(17.4)	(42.9)	(57.2)	(38.0)
High School	104	6	9	55	15	86	14	289
	(40.1)	(54.5)	(50.0)	(49.1)	(65.2)	(33.0)	(22.2)	(38.7)
College	32	-	2	7	2	17	-	60
	(12.3)		(11.1)	(6.3)	(8.7)	(6.5)		(8.0)
PG & Professional	28	-	2	4	-	12	-	46
	(10.8)		(11.1)	(3.6)		(4.6)		(6.2)
Total	**259**	**11**	**18**	**112**	**23**	**261**	**63**	**747**
	(100)	**(100)**	**(100)**	**(100)**	**(100)**	**(100)**	**(100)**	**(10C)**

Figures in parentheses are percentages.

Table 2.16.2: Literacy Status of Various Caste Members (Female)

Literacy	Rajulu	Kapu	Golla	Settibalija	Mangali	Mala	Madiga	Total
Illiterate	26	2	2	11	5	54	19	119
	(10.3)	(25.0)	(11.1)	(11.9)	(22.7)	(21.3)	(32.8)	(16.9)
Primary	104	2	6	48	6	136	31	333
	(41.3)	(25.0)	(33.3)	(52.2)	(27.3)	(53.8)	(53.4)	(47.4)
High School	99	4	9	30	7	52	8	209
	(39.3)	(50.0)	(50.0)	(32.6)	(31.8)	(20.6)	(13.8)	(29.7)
College	19	-	1	3	4	10	-	37
	(7.5)		(5.6)	(3.3)	(18.2)	(3.9)		(5.3)
PG & Professional	4	-	-	-	-	1	-	5
	(1.6)					(0.4)		(0.7)
Total	**252**	**8**	**18**	**92**	**22**	**253**	**58**	**703**
	(100)	**(100)**	**(100)**	**(100)**	**(100)**	**(100)**	**(100)**	**(100)**

Figures in parentheses are percentages.

Table 2.17.1: Details of Occupational Structure among Males

Occupation	Rajulu	Kapu	Golla	Settibalija	Mangali	Mala	Madiga	Total
Farmrelated	124 (65.2)	4 (66.6)	4 (33.3)	48 (68.6)	-	126 (83.4)	25 (83.3)	331 (68.8)
Non-farm	34 (17.8)	2 (33.3)	8 (66.7)	18 (25.7)	18 (81.8)	11 (7.3)	5 (16.7)	96 (20)
Services	32 (16.8)	-	-	4 (5.7)	4 (18.1)	14 (9.3)	-	54 (11.2)
Total	**190 (100)**	**6 (100)**	**12 (100)**	**70 (100)**	**22 (100)**	**151 (100)**	**30 (100)**	**481 (100)**

Figures in parentheses are percentages.

Table 2.17.2: Details of Occupational Structure among Females

Occupation	Rajulu	Kapu	Golla	Settibalija	Mangali	Mala	Madiga	Total
Farm related	-	-	-	22	-	86	26	134
				(91.6)		(81.9)	(96.2)	(81.2)
Non-farm	-	-	-	-	8	12	-	20
					(100)	(11.4)		(12.1)
Services	-	-	1	2	-	7	1	11
			(100)	(8.4)		(6.7)	(3.8)	(6.7)
Total	-	-	**1**	**24**	**8**	**105**	**27**	**165**
			(100)	**(100)**	**(100)**	**(100)**	**(100)**	**(100)**

Figures in parenthesis are percentages.

Literacy

Education is considered as one of the indicators of development. In Godilanka 77% are literate. Among the illiterates (13%) females outnumber the males. Among the literate, primary educated (67%) predominate over those who had high school (21.4%) and college and above level (11.3%) education. Gender differences are clearly noticeable in the educational categories above primary education.

For college education, the villagers have to go to Amalapuram. There is an inverse relationship between the caste rank and illiteracy. Conversely, in the higher echelons of education upper castes predominate over the lower castes (Tables 2.16.1 and 2.16.2).

Occupational Structure

Traditionally caste and occupation were closely related, especially so in the rural areas, in India. There existed an overlap between caste rank and economic status (class). The upper castes were usually better off and lower castes were mostly poor. Thus, 'ritually polluting' castes (ex-untouchable castes) were always involved in menial jobs and agricultural labour. They never possessed land and *Pucca* houses. In contrast, the upper caste groups usually were never engaged in manual labour and mostly lived in better houses. The castes in between were always in a state of flux. The condition today in Godilanka is no different. There continues to be an overlap between caste and economic (also social) class.

Since Rajulu own maximum extent of land, both under agriculture and aquaculture, they get their lands cultivated through Settibalija, who are either owner cultivators or cultivate leased in agricultural lands (tenants) of Rajulu. The scheduled caste members are engaged as labour in agricultural production and aquaculture. Women from Settibalija, Mala and Madiga work as agricultural labour or as domestic servants (Tables 2.16.1 and 2.16.2).

In addition to agriculture, individuals from different castes are engaged in various other activities, both within and outside the village, to eke out livelihood. They are employed in either government or private institutions as employees, engaged in petty trade and business, rendering health services, and other errands within or outside the village (Tables 2.17.1 and 2.17.2).

Table 2.18: Annual Income of Different Households

Annual Income (in '000)	Rajulu	Kapu	Golla	Settibalija	Mangali	Mala	Madiga	Total
Upto 12	-	1	1	2	4	8	5	21
		(16.6)	(14.3)	(4.4)	(36.4)	(7.4)	(19.3)	(6.5)
12-25	15	4	4	23	5	69	18	138
	(12.4)	(66.7)	(57.1)	(51.1)	(45.4)	(63.3)	(69.2)	(42.5)
25-50	31	1	2	12	2	20	3	7
	(25.6)	(16.6)	(28.6)	(26.7)	(18.2)	(18.3)	(11.6)	(21.8)
50-1Lakh	27	-	-	8	-	12	-	47
	(22.3)			(17.8)	-	(11)	-	(14.5)
> 1Lakh	48	-	-	-	-	-	-	48
	(39.7)	-	-	-	-	-	-	(14.7)
Total	**121**	**6**	**7**	**45**	**11**	**109**	**26**	**325**
	(100)	**(100)**	**(100)**	**(100)**	**(100)**	**(100)**	**(100)**	**(100)**

Figures in parenthesis are percentages

Annual Income

Income determines/defines the level of living and, to an extent, interrelationships. While owner-cultivators and tenants obtain their income from agriculture and coconut gardens, aquafarmers earn their income from aqua farms and agricultural labourers through the wages. The annual household income of various caste groups is grouped into five income groups.

Caste and Class polarisation are very clear from the above table. Rajulu have incomes above Rs.12,000/- only. Amongst them a considerable ratio of them are found in Rs. 25-50,000/- and Rs. 50,000-1 Lakh income groups. In the case of others, a majority of them are in the income range of Rs. 12 -25,000/-. Those who have a household income above Rs. 1 lakh are all aquafarmers. It is important to note that there are a few families among Settibalija and Mala in the income range between 50,000-1 lakh. They are all government employees (Table 2.18).

Power Structure in the Village

The ability or authority of an individual or a group to carry out the wishes or policies and to control/coerce or dominate over the behaviour of others, whether they deserve to compete or not, is considered and defined as power. The power is vested either in an individual or a group, who use resources as a means to force their will on others and the basis of the resources lies in social relationships, economic status and the individual's position in a group or society (Bose, 1984).

Power is a social relationship and is often spoken of as a reciprocal relationship. An individual who exerts power over another does not necessarily escape obligations, commitments and limits to his own freedom of action as a result of his power relationship. Consequently, power structure refers to a total network of social and economic relations within a community (Theordorson and Theordorson, 1970). The power structure subsumes not only those who are considered as official leaders in recognised political structures but also those widely accepted influential groups and associations, viz., caste councils within the village (Yogendra Singh, 1969; Leach, 1960).

The basis of power is located in the ownership of land and economic wellbeing (Cohn, 1965, Bailey, 1957). With the passage of time, coupled with the change in economic and social system, the power in the Village Panchayats still continued to be a tool in the hands of landlords. Though the caste councils represented a parallel power structure, they were subjugated to the power of landlords (Singh, 1969: 717). The power centres in Godilanka are: 1) the Village Panchayat, 2) the Primary Agricultural Co-operative Society (PACS), 3) Village Accountant and 4) Caste *Sanghams* (councils).

The Panchayat Office is located in the central part of the village. The Office of the Panchayat comprises a Sarpanch, Upa-Sarpanch and nine members elected for a term of five years. Panchayat elections took place in 1995 after the incorporation of the 73rd Constitutional Amendment[4]. In line with the 73rd Constitutional Amendment, Godilanka Sarpanch position was reserved for Backward Castes. The Sarpanch of Godilanka, who belonged to Settibalija caste, was elected unopposed. The Upa-Sarpanch is a woman from Rajulu caste. Among the nine members elected, three members each from Rajulu and Scheduled Caste, and one member each from Settibalija, Mangali, and Golla represent the Panchayat. Among the scheduled caste representatives in the Panchayat, two are from Mala and one is from Madiga castes. As per the requirement of the Panchayati Raj Act, 1994, 33% of the members should be women. Thus, Godilanka Panchayat has totally three women members, from Rajulu, Settibalija and Mala castes. The level of education among the Panchayat members correlates with the caste ranking in the village. The oldest member in the Panchayat is 62 years old and the youngest member is 33 years old.

The main source of income for the Panchayat is collection of house taxes. The village Panchayat possesses coconut trees, which are grown on the bunds of village tanks and *Kousiki*. Annually, the village tanks are leased-out to the fishermen community of the neighbouring villages. They rear fish in the village tank. For the coconut trees and fish rearing, an open auction is held by the local Panchayat Office under the authority of Executive Officer of the Mandal. These finances collected are source of additional income to the village Panchayat. The Panchayat maintains a library, which subscribes Telugu newspapers.

Village Administrative Officer

The Village Administrative Officer in Godilanka is from Rajulu caste, who had inherited this office from his father. The Village

Administrative Officer, formerly known as *munsab*, maintains village records and collects land revenue and water cess from the farmers for utilising water from the irrigation canals. He is the representative of the government at the village level. His position in the village is quite important. He co-ordinates between the villagers and the revenue department during the times of calamities, elections, in sanctioning land *pattas*, issuing of certificates, ration cards, implementation of development programmes, etc. There are two village servants (*Grama Naukar*), one Chakali (washerman) from Godi and one Mangali from Godilanka. They pass on the information from VAO to the villagers.

Caste *Sanghams* (Councils)

Caste *sanghams* regulate the behaviour of the members of the caste on ritual, social, and political aspects. Caste *sanghams* continue to play an important role in the village social organisation. They work as an organ for promoting solidarity of the caste and for preserving discipline among its members with certain ritual, local administrative, and political authority (Gupta, 1984). Commenting on the nature of caste, Dumont (1957) and Ghurye (1957) considered caste Panchayat as quasi-independent/quasi-autonomous.

In Godilanka, each caste group has its caste *sangham*. They have the authority of imposing fines. The caste *sanghams* are observed to be quite active among the Settibalija and the Scheduled Castes. However, the caste *sangham* of Rajulu is not very active in Godilanka.

Settibalija *Sangham*

The *Sangham* has an overwhelming control on all the moral and social behaviour of the members of Settibalija caste in the village. Members are elected to this *Sangham* and they are known as *'peddamanushulu'*. The *peddamanushulu* are six in number and mediate in all matters concerning their caste members. Issues such as conflict between individuals, brothers/relatives connected with inheritance of property, violation of caste norms, extra-marital relations, etc., of their caste are dealt with by the Settibalija *Sangham*. Meetings of caste *Sangham* are conducted as and when required. In general, meetings are conducted in the night after dinner at Kanakadurga temple at Vasarlagudem to facilitate members to attend the meetings. A nominal amount is collected for common expenses relating to the caste, during festivals, issues relating to common good of the community.

Mala *Sangham*

It consists of Malas from Linepet, Ballavaripeta, Penumalapeta, and Pathuru. There are eight *peddamanushulu* in this *sangham*. They are elected by consensus. The term of office is for one year. In practice, the *peddamanushulu* are middle aged, educated and articulate individuals who have good relations with other caste groups within the village. Recently they tried to negotiate with the farmers and landlords for enhancement of wages to the agricultural labour.

Madiga *Sangham*

Earlier though they had differences with their neighbouring caste Mala, they were part of Mala *Sangham*. With the political situation in the State, on the division of reservation of SCs, had compelled them to separate. They are numerically less but organised into a *Sangham* with three council members. This *Sangham* for advice and direction approached the neighbouring village members of their caste.

Among Golla, Mangali and Kapu there are no formal caste *Sanghams*, as they are numerically small in size. However, *peddamanushulu* of these castes are recognised depending on their status in the village. They act as arbitrators for their respective caste groups in case of disputes or any matters concerning them.

Among Rajulu, the caste *Sangham* decides on the matters of their group and village. They usually comprised big landlords, and presently this is not observed explicitly. However, big landlords are considered as *peddamanushulu*. They control and steer their caste group in the village. The *peddamanushulu* maintain their lobby and groups. At the time of crisis pertaining to their members and their identity, they unite for the general good of their caste, setting aside their personal choices and preferences. The *peddamanushulu* are about 7-10 members and village Panchayat is the place of meeting(s). Indulgence and affiliation to political parties has weakened the strength of their caste *sangham*.

Livestock

The possession of livestock mainly depends on the extent of landholding. Rajulu possesses more than half of the livestock in the village. Among other castes Mala have the most of the buffaloes. This is because after aquaculture many landowners have disposed off their cattle and the Mala caste members started maintaining cattle for using

Table 2.19: Livestock of Different Categories of Residents

Livestock	Rajulu		Kapu		Golla		Settibalija		Mala		Madiga		Total	
	HH	L	HH	L	HH	L	HH	L	HH	L	HH	L	HH	L
Buffaloes	55	74	1	2	2	5	12	23	21	28	4	7	95	139
Cows	12	17	-	-	1	1	1	1	4	4	1	1	19	24
Bullocks	29	95	-	-	1	2	4	8	5	10	1	2	39	117

Table 2.20: Material Possessions of Different Caste Members

Item	Rajulu	Kapu	Golla	Settibalija	Mangali	Mala	Madiga	Total
Television	36(83.7)	-	-	2 (4.6)	1 (2.3)	4(9.3)	-	43 (100)
Refrigerator	25 (100)	-	-	-	-	-	-	25(100)
Radio	58(58.1)	2(1.7)	1 (0.8)	15(12.8)	3 (2.5)	24(20.5)	4(3.4)	117(100)
Fan	75(77.3)	-	2 (2.1)	5 (5.1)	3 (3.1)	12(12.4)	-	97(100)
Cycle	62 (39)	3(1.9)	4 (2.5)	16 (10)	13 (8.2)	46(28.9)	12(7.5)	159(100)
Automobile	69(85.1)	1(1.2)	-	2 (2.4)	-	9 (11.1)	-	81(100)
Car	10 (100)	-	-	-	-	-	-	10(100)
Telephone	24 (100)	-	-	-	-	-	-	24(100)
Tape recorder	84(74.3)	-	2 (1.7)	8 (7.1)	-	19(16.8)	-	113(100)
Sewing machine	-	-	-	5 (38.5)	8 (61.5)	-	-	13(100)
Bullock cart	38(82.6)	1(2.2)	1(2.2)	1(2.2)	-	5(10.8)	-	46(100)

Figures in parentheses are percentages.

them for *araka* (ploughing) on agricultural lands as well as coconut gardens. Among milch animals, Rajulu landowners have most of the cows and the better-off possess them (Table 2.19).

Material Possession

The material possessions observed within a household reflect the class character of that family. The higher the caste and class the higher is the chance of owning them. The material possessions can be broadly grouped as essentials and status goods. Radio, cycle are necessary items whereas possession of colour TV, refrigerator, automobiles, cars and telephone are indicators of higher status. These items are mostly noticed with affluence. Rajulu possess almost all the refrigerators, TVs and automobiles. However, the other caste members own only cycles and tape recorders, even here Rajulu own maximum of them. The lowest in the class and caste in the village do not possess any material possessions (Table 2.20).

The above observations reveal that there is an overlapping of caste and class hierarchies in Godilanka. Rajulu are on the top of the caste and class hierarchy and the Scheduled castes (Mala and Madiga) are on the lower end of the caste and class hierarchy as most of them are employed as labourers. In the intermediary category Kapu, Golla and Settibalija are found. They are the middle caste categories and most of them are tenants of different types. Above the SCs we find the servicing caste Mangali. Thus the agrarian hierarchy follows the caste hierarchy.

NOTES

1. Taluk are the administrative divisions prior to 1984. In 1984 for administrative convenience each erstwhile taluk was divided into two or three Mandals.
2. Konaseema etymologically, in Telugu, is an area at a corner near the sea. This is fertile and the yields are high. It is also known as central delta.
3. Kousiki is a drainage canal, which drains the water from agricultural lands into the sea, which is at a distance of three kilometres from the village. The water in it rises and recedes according to the movement of backwater.
4. 73rd Amendment requires all the state governments to introduce legislative measures for revitalising village Panchayats. It includes mandatory Panchayat elections at regular intervals, reservation of Panchayat seats for women, SC and ST, and Backward castes and devolution of government responsibilities to the village level.

CHAPTER 3

Agrarian Social Structure

This chapter provides a comprehensive overview of the agricultural situation in Godilanka. Agriculture involves categories such as owners, tenants and agricultural labourers, and their relations that are part of the agrarian social system. Depending on the size of the land owned, both in agriculture and coconut gardens, the landowners are classified into different categories. The labour employed by the landowners includes casual ('private') hereditary (*sonthamanishi* and *paleru*) and contracted (tied labourers as *sonthamanishi* and *paleru* for a crop season). This chapter discusses their working conditions, payments, etc. The category of tenants, both *paalu* and *gutta*, have different terms and conditions, working relationships and mutual obligations with the landowner. An attempt is made to understand the interrelationships among these different categories in the present chapter. In addition to these, the role of women in agriculture, credit, which is an essential element of agrarian system in the village structure, etc., are also dealt in this chapter for a better understanding of the agrarian situation in the study village.

I

The economy of Godilanka is mainly agricultural and hence, the relations of production consist essentially of relation between categories of persons contributing in different ways to the processes of agriculture. These categories include landowners, tenants and agricultural labourers. The interrelations, which form into a composite whole, constitute the agrarian class structure of the village. With the coming of modern institutions, like cash economy, mechanisation, modern education, etc., the traditional relationships tended to acquire a new and altered form, which is more contractual in nature and impersonal in character.

The traditional occupations were generally caste-specific and hereditary. Many a new occupation, which cut across the traditional caste barriers, has emerged in recent years. A number of villagers are thus engaged in occupations either in government or private jobs, skilled as well as unskilled, in different cities and towns. Agriculture is the mainstay of the villagers and most of it is oriented to market. The traditional economic activity of the village included the activities of village artisans and services.

In addition to agriculture, coconut gardens are valued high and preferred. This is because, though the initial investment is higher, annual maintenance costs are lower and the returns are relatively higher. That is why the land under coconut garden account for 44.6%, while the land under agriculture is only 36.4%.

Besides coconut gardens and agriculture, a significant proportion of land is under aquaculture (19%) and most of which was under agriculture before aquaculture was taken up. Coconut gardens and agriculture provide employment to a majority of the villagers. These production systems have different sets of economic relations interwoven with one another. Some of them are traditional arrangements, while the others have evolved due to changes that were introduced in the cropping patterns. Thus, the traditional and non-traditional elements in the village economy cannot be straightway compartmentalised.

The caste composition of the village is distinctively dominated by landowners from Rajulu caste. They are economically, ritually and politically dominant, though not numerically. Small landowners and tenants are mostly from Rajulu, Settibalija, and Mala castes. Most of the agricultural labourers are landless and a majority of them belong to the Scheduled Castes, and among the Scheduled Castes specifically they belong to the Madiga caste group.

The class structure of the village reflects the class structure of the region, and to a large extent is conditioned by it. Relations between landowners, tenants and agricultural labourers of the village correspond to such relations in the region as a whole, although each village has its distinctive features. An important aspect over the last few decades has been the changes that have been taking place in agrarian relations due to many reasons. Recent changes in the agrarian relations can be accounted for by the modifications in the relations of production brought about by aquaculture resulting in the return of many villagers, who included both landowners as well as landless labourers.

The system of production in Godilanka has become more complex than it was in the past. The mutual relations of land owners, tenants and agricultural labourers are altered to some extent due to a number of factors such as market forces, introduction of capital intensive agricultural practices, constitutional safeguards for the weaker sections, etc. This, for instance, can be seen in case of Green Revolution, which has resulted in out-migration of members from different castes from the village, especially male members from Rajulu, Settibalija and Scheduled Castes, in search of employment in secondary sector.

II

Andhra Pradesh is one of the important rice-growing regions of southern India. Physiographically, the State comprises three socio-cultural and political regions, viz., Coastal, Telangana and Rayalaseema. The coastal region is the most fertile and irrigated tract, covering almost one third of the area of the state. East Godavari is one of the prominent irrigated districts of the coastal region. River Godavari irrigates this district. In this area two crops of paddy are grown annually (Kharif and Rabi).

In Godilanka, paddy is grown as first and second crop. During Kharif season, the main crop, called *Tholakari* or *Sarava*, is cultivated from June to November. The second crop *dalva* is cultivated from December to April. A third crop, usually green manure/pulses/cattle feed, are cultivated after the harvest of *dalva* and these are raised to enrich the soil. All the agricultural land in the village is irrigated. The coconut gardens, which occupy a considerable extent of the area under agriculture in the village, are considered as un-irrigated. Though according to revenue records coconut gardens are regarded as un-irrigated under Kharif, our study treats it as an annual crop because they yield throughout the year (Table 3.1).

The agricultural season starts in the last week of May or first week of June, i.e., prior to the release of Godavari water into the irrigation canals. After tilling, the fields are kept ready for agricultural operations. The principal stages of paddy cultivation are preparation of nurseries and preparation of the field by ploughing, manuring, transplanting, weeding and harvesting. Demand for agricultural labour will be high during these agricultural operations, although there is a general demand for casual labour, which is mostly provided by the lower caste groups. The intermediary caste groups and SC are mostly dependent on casual labour.

Table 3.1: Crop Seasons in Godilanka

Crop	Period	Irrigated
Tholakari/Sarava (First)	June-November	Paddy
Dalva (Second)	December-April	Paddy
Third*	April-May	Black-gram, green-gram and green manure (*Rotta*)
Annual (Un-irrigated)	Coconut Gardens	

* The third crop thrives on the dampness of the second crop.

Process of Agriculture

After the fields are tilled and levelled with the irrigation water coming into canals starts the agricultural season, which is around the last week of May and first week of June for *tholakari* and first half of November for *dalva.*

On an auspicious day the sowing of seeds takes place. The elders of the family decide the date in consultation with a Brahmin priest. Usually the farmers refer the local *panchangam* (almanac) and follow *karthulu.* In the field after a brief chanting of *mantras* (hymns) and performance of *pooja* (ritual) a coconut is offered to the God in anticipation of a good harvest. After the performance of the ritual they broadcast the seeds. The first seeds are placed in the nursery in the north-east corner (*ishanyam moola*) of the field, which is believed to bring prosperity.

For every agricultural activity, both at the time of initiation as well as at the end of the activity, offerings to God(s) and ancestral spirits is an important feature, which is performed usually by breaking a coconut at the inception of an agricultural operation. This is done to appease the gods and spirits to give them a bounty of crop.

Among Rajulu, agriculture is totally a male-centred activity. The elderly males who are believed to be lucky in their family or among their relatives sow the first seeds in the nursery. Among the other castes practising agriculture, it is women who take active part in sowing the first seeds in the fields. In general, in these caste groups' women give offerings with reverence to the God and continue the sowing. However, menstruating women are forbidden to take part in the ceremony or in the agricultural work.

The cultivators prepare nurseries and they are made in a corner of one's agricultural field. For this the land will be tilled thoroughly and watered by the labourers from Settibalija caste and SCs. The seeds are sown and the seedlings, after they grow to a desirable size, are transplanted by hired female agricultural labour mainly from the SCs. After a period of time chemical fertilisers are applied to the fields. Weeding operations are undertaken after three or four weeks by female labourers. Again, chemical fertiliser will be applied to the crop. With continuous supply of water to the fields through canals, the crop grows to an optimum size. At this stage, in order to protect the crop from pests/ insects, pesticides or insecticides are sprayed. Later, finally, the crop is harvested. To harvest the crop, mostly women are employed. A few men are employed to stack the harvested crop.

Table 3.2: Land under Various Crops by Season (in Acres)

Year	Kharif (June-Nov)	Rabi (Nov-Apr)	Third crop (Apr-May)	Annual crop (Coconut garden)
1991-92	512	490	50	375
1992-93	510	510	70	375
1993-94	443	440	60	375
1994-95	391	385	75	375
1995-96	381	380	60	375
1996-97	355	360	80	373
1997-98	351	351	55	364

Source: MRO, Allavaram.

It is important to note that the lands under different crops show two distinct trends. By and large, land under un-irrigated remained more or less constant from 1991 to 1998. In contrast, the land under irrigated agriculture has come gradually over years. This gradual decline is a result of the introduction and expansion of aquaculture. Thus, between

93-94 and 94-95 there is a shift from irrigated agriculture to aquaculture (Table 3.2).

III

Production Relations

Production is a socially organised activity. This involves participation of different types of persons engaged in production. The interrelation and involvement tends to bind people with special rights, duties, and obligations in the process of production. The relationships of production distinguish persons in terms of castes as well as classes. In an agrarian society, the basic differences among the residents depend on the possession of agricultural land. In other words, it is the ownership and the non-ownership of land, which matters in defining the production relationships.

Caste and class are the two important categories to study the agrarian social structure. Béteille (1969) viewed that the caste and class relations are important and have certain common properties. He feels that the study of class has to deal with two sets of relationships: the intricacy of distribution (ownership) and of the interrelationship between landowners and non-landowners who are part of production system. He developed the concept of agrarian hierarchy. This hierarchy is normally distinguished by the category of landowners, tenants and agricultural labourers (Chart 3.1).

Landowners

The category of owners, tenants and agricultural labourers is not homogeneous. As noticed in the previous chapter, most of the landowners are from Rajulu caste, while all others have only a small proportion of land (Table 2.14). The landowners in Godilanka are classified into five categories based on the extent of land that they own: 1) marginal farmers, 2) small farmers, 3) medium farmers, 4) large farmers and 5) big farmers. Marginal farmers are those who possess agricultural land below one acre. Small farmers own land between an acre and 2.5 acres. Medium Farmers have agricultural land, which is above 2.5 acres and below five acres. If the land holdings are more than five acres and less than ten acres, they are categorised as large farmers. All those farmers who possess land above ten acres are considered as big farmers.

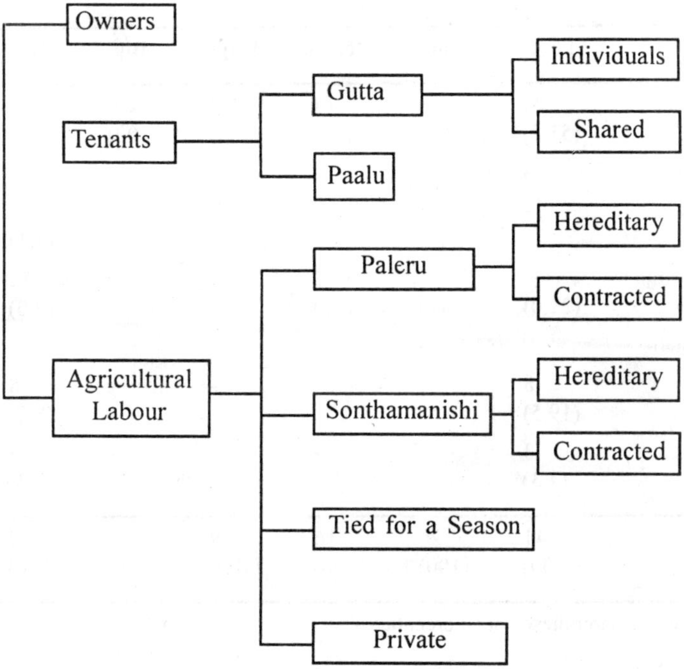

CHART 3.1: SOCIAL ORGANISATION OF AGRICULTURE

If one looks at the landholding categories, most of the Rajulu belong to small and medium categories. The large and big farmers among them if put together are more than the medium farmers. In contrast to Rajulu, among the Settibalija, Mala and Madiga marginal farmers form the largest category among them. In fact, among both Mala and Madiga all the landowners are marginal farmers only. Among Golla there is only one person who owned land belongs to small farmer category. The caste groups Kapu and Mangali do not possess any land (Table 3.3 on next page).

It is pertinent to note that at the upper levels of the agrarian hierarchy there is a greater concentration of land among less number of people. In contrast, in the lower levels of the agrarian hierarchy there is less amount of land among majority of the people. The agricultural land possessed by the big landowners (43%) is more than the land possessed by the marginal, small and medium farmers (37%) together (Table 3.3.1).

Table 3.3: Category-wise Agricultural Landowners

Caste	Marginal	Small	Medium	Large	Big	Total
Rajulu	21	18	14	9	6	68
	(51.2)	(90)	(88)	(100)	(100)	(73.9)
Kapu	-	-	-	-	-	-
Golla	-	1	-	-	-	1
		(5)				(1.1)
Settibalija	9	1	2	-	-	12
	(22.0)	(5)	(12)			(13.0)
Mangali	-	-	-	-	-	-
Mala	8	-	-	-	-	8
	(19.5)				-	(8.7)
Madiga	3	-	-	-	-	3
	(7.3)					(3.3)
Total	41	20	16	9	6	92
	(100)	(100)	(100)	(100)	(100)	(100)

Figures in parentheses are percentages.

Table 3.3.1: Land Ownership of Caste and Agrarian Class (in acres)

Caste	Marginal	Small	Medium	Large	Big	Total
Rajulu	20.5(0.97)	35.6(1.97)	44(3.14)	70(7.78)	151(25.2)	321.1(4.72)
Kapu	-	-	-	-	-	-
Golla	-	2(2)	-	-	-	2(2)
Settibalija	8.5(0.94)	2.4(2.4)	8(4)	-	-	18.9(1.58)
Mala	7(0.87)	-	-	-	-	7(0.87)
Madiga	2(0.67)	-	-	-	-	2(0.67)
Total	38(0.92)	40(2)	52(3.25)	70(7.78)	151(25.2)	351(3.82)

Figures in parentheses are averages of that category.

The agrarian hierarchy, by and large, coincides with the caste hierarchy. Even within each category the caste differentiation can be seen in terms of land ownership. For instance, among the marginal farmers, the average landholding differs between Rajulu and the SCs; between Rajulu and other middle ranking castes (Table 3.3.1). Thus it is clear that even the class hierarchies follow the caste hierarchies.

Land and its relations of production in this village have evolved over a period of time. It is very essential to have some awareness of this because of its influence on the contemporary agrarian relation. The development of this village has a close relationship with the Rajulu caste and among them the hold of certain lineages on the landholding is quite pertinent to be observed. The original village was near the Pathuru tank, where only *Rudraraju* lineage members from Rajulu caste used to reside. Among them, those who had only daughters got them married to persons from other villages who moved to this village after marriage to look after the property of their parents-in-law. Similarly, there are many others who moved into the village for different reasons. Thus the lineage members from *Alluri*, *Sagi*, *SayyapuRaju*, etc., have moved into Godilanka. The following few cases illustrate the role of kinship in the control and transfer of land.

Chilvuri Krishnam Raju, aged about 70 years, has two daughters and a son. His second daughter after giving birth to a child (boy) died. The deceased woman's husband married another person. The deceased woman's brother, Sanyasi Raju (son of Krishnam Raju), and the boy's *menamama* (mother's brother) took care of him. The boy, Narasimha Raju, belonged to *Sayyapuraju* lineage. After educating the boy, Sanyasi Raju gave his daughter in marriage to Narasimha Raju and also gave him in dowry part of his landed property. Thus *Sayyapuraju* Narasimha Raju became a resident of this village.

Sagiraju Subba Raju, who is about 50 years, came to Godilanka from Pathepuram village from the neighbouring West Godavari district. He is referred to as 'Pathepuram Raju'. Godilanka was his grandmother's place. He had some conflict with agricultural labourers due to some illicit relationship in which he was involved. He was caught red-handed and this led to a loss of face to him. After this incident no one came to work on his lands in his natal village. Thus, he came to Godilanka, his maternal grandmother's place, and bought some agricultural land and coconut garden and started residing in this village.

Pericherla Rakesh Raju was a farmer in Vedurulanka (15 kilometres from Godilanka) had five acres of paddy cultivation and maintained attached labourers in his fields. He was continuously pressurising the labourers working with him to support and vote for the candidate of his choice in the local Panchayat elections and the labourers were little interested in the candidate of their landlord. In the elections the candidate supported by Rakesh Raju had lost by a narrow margin. From then on Rakesh Raju started harassing his labourers in one form or the other. One day he beat one of the labourers. This issue was taken to the village Panchayat. After hearing the matter, the village elders have felt that it was not appropriate on the part of Rakesh Raju to have beaten the labourer. The village elders had asked him to apologise to the labourer. He did not abide by the decision of the village elders as he felt that this would belittle him. Consequently, the labourers refused to work in his fields. Under these circumstances, he sold his land in that village and moved to Godilanka, his maternal uncle's village, where he purchased agricultural land and settled.

Nadimpalli Rama Raju's father, Simhadri Raju, was originally a native of Komaragiripatnam, which is about 2 km away from this village (VarahalaRaju, 1980:87). Rama Raju's mother was the only child to *Rudraraju* Lakshman Raju, a resident of Godilanka. After marriage, Simhadri Raju came and settled in Godilanka to look after the property of his father-in-law, as his wife was the only daughter. The entire property of his father-in-law was later transferred to him. Following this, he disposed off his share of property at Komaragiripatnam and reinvested that money in Godilanka. At present, Rama Raju, who is around 65, has about 90 acres of both coconut garden and agricultural land and is one of the richest persons in the village. He also has a considerable amount of aqua land.

A majority of landowners in Godilanka possess both agricultural land as well as coconut gardens. Coconut gardens are considered as more valuable than paddy lands as they provide income every month. Besides, the investment, persons employed, owner-labourer relations also differ between them.

Coconut Gardens

Coconut crop is a dry crop and the coconut trees were raised in fields that were specially prepared for this purpose a few decades ago. As compared to paddy cultivation, coconut gardens require less attention.

After the plantation of coconut saplings, it takes five to six years for the first harvest. The harvest continues for quite a long period and most of it is marketed. Compared to paddy cultivation, coconut garden requires less investment and its average returns per acre are high. However, this also requires physical labour and a few casual labourers are employed on the day of harvest/ploughing/ fertiliser application.

The coconut gardens are seasonally ploughed at intervals of time for better upkeep of the farm. A wooden plough is used for ploughing. Farmyard manure and chemical fertilisers are applied twice or thrice annually to maintain the yield of the garden. Progressive and entrepreneurial farmers also improve the coconut gardens by adding a new layer of fresh soil. For taking care of the above activities, bullocks and *sonthamanishi(s)* are maintained. The harvesting of coconuts is done once in a month and the average harvest per tree varies from 75-85 numbers annually.

The ownership of coconut gardens is also heterogeneous as observed in agriculture. A similar classification that was followed in agriculture is also adopted in coconut farming. Depending on the ownership, as in agriculture, coconut farmers are grouped into marginal, small, medium, large and big farmers.

Table 3.4: Categories of Farmers Possessing Coconut Gardens

Caste	Marginal	Small	Medium	Large	Big	Total
Rajulu	32	17	13	7	6	75
	(42.7)	(22.7)	(17.3)	(9.3)	(8)	(100)
Kapu	-	-	-	-	-	-
Golla	1	2	-	-	-	3
	(33.3)	(66.7)				(100)
Settibalija	7	1	1	2	-	11
	(63.6)	(9.1)	(9.1)	(18.2)		(100)
Mangali	-	-	-	-	-	-
Mala	11	3	-	-	-	14
	(78.6)	(21.4)				(100)
Madiga	4	2	-	-	-	6
	(66.7)	(33.3)				(100)
Total	55	25	14	9	6	109
Total	(50.5)	(22.9)	(12.8)	(8.3)	(5.5)	(100)

Figures in parentheses are percentages.

The trend observed in case of agricultural land is also noticed among coconut farmers. In all categories of coconut farmers Rajulu, as in agriculture, predominate all the other castes. Except for Settibalija, all the other middle ranking castes and lower castes have only marginal and small landholdings. Settibalija, Mala and Madiga caste groups have a higher proportion of marginal farmers among them. Only three landowners among Settibalija are medium and large coconut farmers. Thus, there is an overlap between the caste ranking and ownership, and the extent of coconut garden one owns (Table 3.4 and 3.4.1).

Table 3.4.1: Coconut Gardens by Caste and Agrarian Class (in Acres)

Caste	Marginal	Small	Medium	Large	Big	Total
Rajulu	30.5 (0.95)	34.7 (2.04)	48.8 (3.75)	52 (7.42)	154 (25.6)	319.6 (4.26)
Kapu	-	-	-	-	-	-
Golla	1 (1)	4 (2)	-	-	-	5 (1.67)
Settibalija	5.9 (0.84)	2 (2)	3.6 (3.6)	11 (5.5)	-	22.8 (2.07)
Mangali	-	-	-	-	-	-
Mala	5.8 (0.52)	6 (2)	-	-	-	11.8 (0.84)
Madiga	1.8 (0.45)	3.3 (1.65)	-	-	-	4.8 (0.8)
Total	45 (0.82)	50 (2)	52 (3.71)	63 (7)	154 (25.6)	364 (3.34)

Figures in parentheses are averages of that category.

Coconut Gardens and Labour

The characteristic feature of coconut gardens in Godilanka is that they give yield once in a month. As pointed out earlier, the owners of the coconut gardens employ *sonthamanishi* to work with them in these gardens. In addition to the *sonthamanishi*, they engage *dimpu* workers (persons who are adept in climbing the coconut trees and in harvesting coconuts) as well as casual labourers on the harvest days. In Godilanka, the *dimpu* workers are from Madiga caste only. There are a few *dimpu* workers from the neighbouring villages working in Godilanka. They belong to both Settibalija and Madiga castes.

The *dimpu* workers (coconut harvesters) organise themselves in to small groups depending on the size of the coconut garden. The size of this group varies from 3-4 members. There are three such groups working in Godilanka. This type of organising is also beneficial for large and big farmers, as this enables them to engage these groups and get their job accomplished in a few days. They follow the harvest cycle on a regular basis. The average number of days of work in a month for *dimpu* workers varies between 15-20 days and during the rest of the time some work as 'private' labourers. During the coconut harvest days they are paid either daily wages by small and marginal farmers or on a work contract basis by the medium and above category of farmers, depending on the number of acres harvested. In both types a *dimpu* worker earns Rs. 70-80/- in a day. The *dimpu* workers are usually given consumption loan. The medium farmers give a loan amount between Rs.100-300/-, which is interest free and is reclaimed from their wages. However, the large and big farmers advance bigger loans. The loan ensures timely labour. Unlike the other labourers, the *dimpu* workers are not hereditary but contracted only.

Caste has a stranglehold on the land transactions. Any land transactions in Godilanka can not take place without the knowledge of Rajulu. In this regard the role of Village Administrative Officer (VAO) is quite important. The VAO, earlier known as *Munsab*, who hails from Rajulu caste and belongs to *Rudraraju* lineage. He maintains the details of all the land holdings in the village, including ownership title, ownership rights, taxpayers, etc., relating to the land assets. This VAO has been playing a crucial role in the land transactions as the land transactions take place only with his involvement. He was helpful to his community members in furnishing information regarding the lands that come for sale and thus sees to it that the land does not go to other caste/community members.

Some of the titleholders of agricultural land reside outside the village. These people have gone out to work as employees in government or in private sector. Few of them had started doing business in the nearby towns. Also, many Rajulu own lands in other neighbouring villages, such as Rellugadda, Mogallamuru, and Turupulanka. *Nadimpalli* Rama Raju is the biggest landlord in Godilanka and owns about 100 acres of land. There are a few big landlords (6 members) who possess about 40 acres of land, both paddy and coconut gardens.

Some Rajulu landowners in Godilanka, who possessed large tracts of land, retain labour on a regular basis. It was essential for the landowners to maintain the labourers because they need them to work in their fields' continuously. This kind of selective employment of certain set of people over others had necessitated the labourers to continue with the same landowner for a longer period of time and even generations.

IV

Hereditary Relations

In Godilanka the landowners employ agricultural labourers to work in their fields. While engaging them in work the landowners advance consumption loan which is reclaimed in their wages every month. In case of *paleru*, an attached labourer who looks after the cattle, the recovery of consumption loan is done annually as his payments are on an annual basis. Besides consumption loan, for both the above types of labour credit is given as and when required to meet their family needs. The amount of loan advanced if exceeds their wages it will be carried forward to the following season/year. Thus, the above category of labourers, in a way, gets tied to the same landowners and work for several years and even over generations. If these labourers can clear their debt, they would be free to work with any one else. But in practice, it is of rare occurrence. In fact, it is not the *sonthamanishi* or *paleru* alone who are tied to the landowner but their family members also get tied to the landowners. In Godilanka, it is observed that certain lineages of agricultural labourers are engaged with the same land owning families for generations.

The hereditary continuation was also a form of security for the agricultural labourers. This type of arrangement is observed with large and big farmers in Godilanka. About 14 landowners have engaged 23 agricultural labourers in such an arrangement (Table 3.6). The following case of Venkata Raju gives an opportunity to understand the hereditary continuation of relations over generations.

Kasi Venkata Raju's father, a Mala caste member from Linepet and land less labourer, was working as *paleru* with Narasimha Raju's father for a long period and even Venkata Raju worked for a brief period with

Narasimha Raju. Venkata Raju's father had three sons. One of his sons after marriage had left for his in-law's place, and his remaining two sons continued to work in the fields of Narasimha Raju (Chart 3.2).

Landowner **Attached Labour**

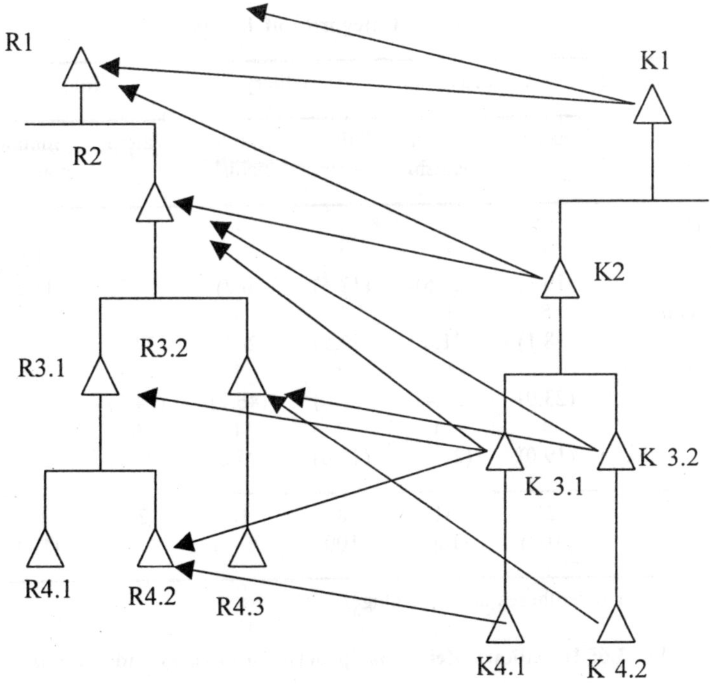

CHART 3.2: HEREDITARY RELATIONSHIP BETWEEN *RUDRARAJU* AND *KASI* LINEAGES

In Godilanka nearly three fourths of attached labour, *paleru* and *sonthamanishi*, is contracted labour. A large numbers of both contracted and hereditary *paleru* and *sonthamanishis* are working with medium, large and big landowners (Table 3.5 on page 90).

Chandra Raju, second son of Narasimha Raju, is referred to as pleader in the village as he was the first person to obtain law degree in Godilanka. He had inherited 30 acres of land from his father. Of this, 18 acres was under coconut and the rest was paddy. Venkata Raju has been working with this family for the past 40 years. He was engaged to work by his father at Narsimha Raju's (pleader's father) house when he was 12 years old. Initially, he was employed to look after the cattle. This he did till he was 18 years old, after which he was engaged to work in the fields as *sonthamanishi.*

Table 3.5: Categories of Labour

Farmer type	Contracted		Hereditary		Total	
	Paleru	Sontha manishi	Paleru	Sontha manishi	Paleru	Sontha manishi
Marginal	-	-	-	-	-	-
Small	4 (19.0)	11 (21.6)	1 (12.5)	1 (6.7)	5 (17.3)	12 (18.2)
Medium	8 (38.1)	11 (21.6)	3 (37.5)	3 (20.0)	11 (37.9)	14 (21.2)
Large	5 (23.9)	15 (29.4)	2 (25.0)	7 (46.7)	7 (24.1)	22 (33.3)
Big	4 (19.0)	14 (27.4)	2 (25.0)	4 (26.6)	6 (20.7)	18 (27.3)
Total	21 (100)	51 (100)	8 (100)	15 (100)	29 (100)	66 (100)

Figures in parentheses are percentages.

Table 3.6: Hereditary Relationship between Owners and Labourers

Name of the family	Lineage of agricultural labour hereditarily working
Alluri T Raju	Akumarthi, Lanka
Alluri V Raju	Pothula, Bommi
Alluri S Raju	Balla
Nadimpalli R Raju	Penumala
Rudraraju K Raju	Gosangi, Kasi
Rudraraju M Raju	Konuku, Settibathula
Rudraraju N Raju	Settibathula
Rudraraju T Raju	Kasi, Settibathula
Rudraraju S Raju	Katari, Ethakota

Narasimha Raju had two sons. In his old age Narasimha Raju had divided his property between his two sons. Along with the division of property, the labour working with him was also distributed accordingly. Thus, Venkata Raju was given to pleader and his elder brother (Venkat Reddy) went to work with pleader's elder brother Rama Raju. Since then, Venkata Raju has been working with the same landowner.

Venkat Reddy has a son, who is at present working with Rama Raju as *sonthamanishi.* Venkata Raju has a son (Srinu) who studied till seventh class and became a tractor driver to the pleader. This particular case shows the manner in which a landowner's family and the family of the tied labourer was enmeshed hereditarily for generations.

V

Tenants

A large proportion of cultivators engaged in farming is non-landowners. They work on the land belonging to others. This constitutes the essence of economic as well as social relations between different caste/classes engaged in agriculture. In the broad category of tenants, there are two distinct types working under different sets of conditions: *paalu vyavasayam* (share cropping) and *gutta vyavasayam* (fixed tenancy). Srivastava (1989) in Uttar Pradesh, Sarap (1998) and Mohapatra (1996) in Orissa, Murty (1987) and Parthasarathy (1991) in Andhra Pradesh have found similar pattern of tenancy. Before dealing with each category of tenants, it will be proper to know who the tenants are and what categories of people are involved in tenancy.

A tenant is one who leases-in land for cultivation from a landowner and in some cases, he can be simultaneously a landowner as well as an individual who leases-in land for agriculture. The category of tenants is not a homogeneous group. A tenant is not fixed for all time with a single landowner. Continuance of tenancy is dependent on the attitude of the landowner in leasing-out his agricultural land to the same individual. Thus the category of tenants is a residual category. The tenancy varies from season to season and if some one is a tenant in one season/ year he may or may not be a tenant in the next season/year. Sharecropping in Godilanka is known as '*paalu vyavasayam*' (hereafter referred to as *paalu*). Locally *paalu* means share. Similarly, contract farming for a fixed amount of produce is known as *gutta.*

Among different types of cultivation in Godilanka, *paalu vyavasayam* is the highest followed by *swanta* (own) and *gutta vyavasayam*. In *swanta vyavasayam* more than three-quarters of landowners are Rajulu. Among Settibalija, more than half of the land leased-in is either under *gutta* or *paalu*. Those who practise *gutta* are considered economically better off than those who practise *paalu vyavasayam*. In case of Settibalija, the practitioners of *gutta* are more than the *paalu*. Mala and Madiga farmers cultivate more than half of the *paalu* land and they practise no *gutta vyavasayam*.

The tenants in Godilanka are either pure tenants without any land or tenants owning some amount of agricultural land (owner-sharecropper or contract farmer). The land under both the above types of tenancy is 255.6 acres. Most of the land is under pure tenants (70.8%) and the rest is under owner-*paalu/gutta* tenants (Table 3.7).

Among those who own land and do *gutta* or *paalu vyavasayam*, a contrasting trend is observed between Rajulu and other castes. In case of Rajulu the amount of land leased-in is lesser than the amount of land owned. Further, there is no *paalu vyavasayam* observed in this category of farmers. In case of the others land leased-in is more than the land owned (Table 3.7).

In Godilanka about three-fourths of the farmers (72.3%) are continuing in *paalu vyavasayam* and the rest (27.7%) are engaged in *gutta vyavasayam*. Most of the Rajulu tenants (60%) are involved in *gutta vyavasayam*, while most of the Settibalija and all the Mala and Madiga tenants are involved in *paalu vyavasayam* only (Table 3.8).

Paalu Vyavasayam (Sharecropping)

According to Byres, "Sharecropping is as old as recorded history"(1983: 7). However, its origins are not precisely known. It can be understood by its occurrence in various forms, which was studied by many researchers from different countries, in various regions (Byres, 1985; Gough, 1981; Mencher, 1978; Thorner and Thorner, 1956; and Kotevsky, 1964). Sharecropping in its economic context is viewed as 'a relationship, a web of dependency' incorporating both economic and non-economic structure of dependency and subordination (Cooper, 1985). Sharecropping is considered as an important element in the agrarian structure of India (Thorner, 1956; Khusro, 1973; and Rao, 1971: 583). Khusro (1973: 126) estimated that about a half of the total cultivated

Table 3.7 Land under different types of Cultivation by Caste

Type	Rajulu	Kapu	Golla	Settibalija	Mala	Madiga	Total
Own	73.2 (77) (50)	-	3 (3) (100)	16.3 (17) (13)	2.9 (3) (5)	-	95.4 (100)
Gutta	29 (37) (20)	2 (2) (100)	-	48 (61) (39)	-	-	79 (100)
Paalu	21 (21) (14)	-	-	24 (23) (20)	44 (43) (70)	13 (13) (100)	102 (100)
Own & gutta	24 (59) (14+10)* (16)	-	-	17 (41) (5+12)* (14)	-	-	41(100)
Own & *paalu*	-	-	-	18 (54) (8+10)* (14)	15.6(46) (6.6+9)* (25)	-	33.6 (100)
Total	147.2(42) (100)	2 (.5) (100)	3 (1) (100)	123.3 (35) (100)	62.5 (18) (100)	13 (4) (100)	351 (100)

* The first value is own land and the second value is either paalu or gutta land leased-in.

land in India was under sharecropping. Sharecropping was more observed in rice producing regions of eastern and southern regions of India and wheat producing areas of the northern regions of India (Khusro, 1973; Rao, 1971: 583; and Ladejinsky, 1965). Sharecropping was known in different areas under different names, viz., *bargadar*, *adhian*, *bhagdan*, etc. While describing sharecropping in western India, Thorner (1956: 46) points out that it is one of the most common arrangements of working on land and it holds good even today.

Table: 3.8 *Paalu* and *Gutta* Farmers by Caste

Caste	*Paalu*	Gutta	Total
Rajulu	12 (40) (11.8)	18 (60) (46.1)	38 (100) (26.9)
Kapu	-	1(100) (2.6)	1 (100) (0.7)
Settibalija	28 (58) (27.5)	20 (42) (51.3)	48 (100) (34.0)
Mala	49 (100) (48.0)	-	49 (100) (34.7)
Madiga	13 (100) (12.7)	-	13 (100) (9.2)
Total	102 (72.3) (100)	39 (27.7) (100)	141 (100) (100)

Figures in parentheses are percentages.

André Béteille (1972: 22) noted the prevalence of sharecropping in the states of West Bengal, Tamil Nadu, and Kerala. According to him, this is because the landowners prefer to avoid manual work, particularly when it is 'arduous' and 'onerous'. This type of attitude is stronger in areas of wet paddy cultivation than in the areas of wheat cultivation.

The prevalence of sharecropping in Andhra Pradesh was widely reported. The study by Parthasarathy and Prasad (1974) which was made in the context of the introduction of new technology in the 1970s (under green revolution), which had altered the sharecropping relationships. They had opined that there was a transition from 'traditional' form to 'cost-share' leasing. The peasants considered this as sharecropping, but there was a qualitative change in the relationships. Sharecropping was regarded as an equal relationship, where in the cultivator and the landlord provide equal inputs, take equal risk and receive equal share of crop.

The produce (grain) after the harvest will be divided between the landowner and the *paalu* farmer. In general, it is observed that small areas of agricultural land are given for *paalu*. Under this, both the landowner and the sharecropper enter into an agreement over the land leased-out. Almost all the landowners from medium and above categories lease-out a small parcel of their land, while they cultivate the rest by employing labour. The landowner provides seeds. After consultation with the landowner, the tenant prepares seedbed for nursery and later transplants the saplings. During the agricultural operations like application of fertiliser and pesticide, harvesting, threshing, etc., the tenant provides labour. The agricultural operations will be looked after either by the landowner or his *sonthamanishi*, while the tenant, who will also work along with the labour employed, will supervise the entire operation. Thus, while the tenant provides the labour input, the landowner supplies agricultural inputs.

The harvested crop is threshed, winnowed and piled at the *kallam* (an elevated portion in the field where the post-harvest activities are carried out). The harvest thus stacked is kept in the *kallam* till the grain is shared between the landowner and *paalu* farmer. The sharing of grain is done only in the presence of the landowner and the tenant has to wait for him. Sometimes the grain remains in the *kallam* for one week till they find an auspicious day for measuring the grain for sharing. In order to protect the piled grain from theft, some patches of wet sand are kept here and there around the stacked grain and impressions of *mudraballa* (wooden insignia specific to each landowner) are stamped on the wet sand patches. Any slight disturbance of the heap of grain disturbs the *mudraballa* impression. This is a precaution against theft or deceit. On an auspicious and convenient day, the landowner and the *paalu* farmer measure and divide the grain according to *paalu* agreement, after deducting the costs of agricultural inputs, such as seed, chemical fertiliser, pesticides, etc., valued in terms of grain equivalent.

The division of produce among the landowner and *paalu* farmer depends on the crop season. For *tholakari*, the produce is divided into three parts, one part for the sharecropper and two for the landowner. For *dalva*, it is divided equally among the landowner and the *paalu* farmer. The by-products of agriculture, such as paddy straw, are mostly retained by the landowner. However, few bundles of straw are given to *paalu* farmer as a consideration by the landowner, if he possess cattle. Decision-making is completely vested in the landowner and this can be clearly observed in case of selection of seeds, type of fertiliser, harvesting time, the day of sharing of the crop, etc., and the *paalu* farmer has to abide by these decisions. The land under *paalu vyavasayam* during 1997-98, both Kharif and Rabi seasons, was 121 acres (Table 3.6).

In Godilanka, there are two types of *paalu* farmers, attached (*sonthamanishi*) and un-attached. Among the *paalu* farmers more than two thirds of them belong to attached *paalu* farmers and all of them have been working as *sonthamanishis*. On the other hand, the un-attached paalu farmers have no such bindings to the landowners as the *sonthamanishis* have. They actually came into existence only after the emergence of aquaculture.

The un-attached *paalu* farmers worked at some time with the landowners or their families as *sonthamanishis*. As substantial number of landowners have turned to aqua farming they had little time to monitor their fields. This shifting to aquaculture had resulted in reduction of work for those agricultural labourers working as *sonthamanishis* with the landowners who turned to aqua farming. These displaced *sonthamanishis* were given land for *paalu* farming without any kind of other obligations.

Most of the *paalu* farmers are from Mala and Madiga caste groups. It is pertinent to note that there is an inverse relationship between the caste status of a *paalu* farmer and the type of *paalu* farming in which they are involved in. The higher the caste rank the lower is the chance to work as *sonthamanishi* and the converse is true with the lower castes in the category of un-attached *paalu* farmers.

All the attached *paalu* farmers from the neighbouring villages are from the middle rank caste groups, like Kapu and Settibalija, and all of them work with large and big landowners (Table 3.9 on page 98). These large and big landowners prefer middle caste *sonthamanishis* to the lower caste persons. This preference is related to rules of purity and pollution.

Contractual Obligations

A landowner keeps some land for self-cultivation and the rest he leases-out to *paalu* farmers. The landowner also needs labour for cultivating his lands. It is obligatory on the part of the sharecropper to provide one person from his (their) family to work as *sonthamanishi* or *paleru* on a contract basis. *Paleru* looks after the cattle and does other errands in the house of landowner, while *sonthamanishi* works in landowner's agricultural fields. This kind of labour is a security for the landowner to carry out his agriculture unhindered. Besides, during peak agricultural operations it is also obligatory for the *paalu* farmer to provide labour from his family, including women. The family members of the *paalu* farmer also work as casual labour during peak agricultural seasons. Unlike the other labourers, they are treated as attached labour and are usually underpaid. This kind of inter-linked labour transaction was found to be more prevalent in irrigated villages than non-irrigated villages and it is always a dominant-dependent relations (Swain, 1999: 2757).

Table 3.9: *Paalu* Farmers by Type and Caste

Caste	Attached farmers	Un-attached farmers	Total
Rajulu	3 (25.0) (3.7)	9 (75.0) (28.1)	12 (100) (10.6)
Settibalija	12 (41.4) (14.8)	17 (58.6) (53.1)	29 (100) (25.7)
Mala	46 (93.8) (56.7)	3 (6.2) (9.4)	49 (100) (43.4)
Madiga	10 (76.9) (12.4)	3 (23.1) (9.4)	13 (100) (11.5)
Neighbouring village	10 (100) (12.4)	-	10 (100) (8.8)
Total	81 (71.7) (100)	32 (28.3) (100)	113 (100) (100)

Figures in parentheses are percentages.

In case of coconut gardens, a landowner also needs labour on the harvest days, which will be carried out on a regular basis for a few days in a month. The number of days of harvest depends on the size of the farms. The *paalu* farmer during this time has to work for the landowner as *sonthamanishi*, and he gets wages lower than the prevailing wage rate in the region. At times, when a *paalu* farmer is unable to attend this work, he has to send any male member from his family or arrange a hired labour. In such an event, the landowner pays the wage of a *sonthamanishi* to the hired labour and the differential wage has to be paid by the *paalu* farmer to the hired labour.

Maintenance of Relations

In order to continue *paalu vyavasayam* uninterruptedly with the same landowner it is essential for the *paalu* farmer to maintain both economic and social relations with him. The economic compulsion and absence of alternative employment opportunities force the *paalu* farmer to work under the landowner. Usually, a *paalu* farmer arranges agricultural labour for the landowner and it is below his dignity to work as agricultural labour.

It is quite common in Godilanka that *paalu* farmers tend to continue with the same landowner for several years. There are also cases of *paalu* farmers who had continued for more than 10 years with the same landowner. There are about 10 cases of Settibalija *paalu* farmers who had been working with the same Rajulu landowners for about 15 years and above. There are also two Settibalija families who are continuing as *paalu* farmers for the second generation.

This can be clearly understood from the case of Vasarla Rama Rao, a 64 years old *paalu* farmer from Settibalija caste. He is a resident of vasarlagudem and along with his elder brother worked with *Rudraraju* Subba Raju's family for about 30 years as *sonthamanishi*. In lieu of the labour provided by Rama Rao, Subba Raju leased out five acres of agricultural land to Rama Rao for *paalu vyavasayam*. After the demise of Subba Raju, Rama Rao worked with his son *Rudraraju* Ramakrishna Raju (52 years) for about 9 years. Later, Rama Rao's sons, Suresh (36) and Naresh (32), continued to work with Ramakrishna Raju on the same land. Before the introduction of aquaculture, they were doing *paalu vyavasayam* in 5 acres of agricultural land and it has come down to 3 acres after the introduction of aquaculture. Earlier to introduction of aquaculture in Godilanka, many Settibalija families and some landless Rajulu did *paalu vyavasayam*. The area under agriculture in Godilanka has come down owing to aqua farming, thus partly disrupting relations that existed for generations between the landowners and *paalu* farmers.

Exchange Labour in Paalu Vyavasayam

Paalu farmers during peak agricultural operations exchange labour among themselves. A *paalu* farmer along with his family members works in his fellow *paalu* farmer's field and they in turn also work for equal number of days for him. The exchange of labour is always between the members/families of the same caste group. This is more frequently observed among the Mala and Madiga caste groups and to a certain extent, among the Settibalija. Exchange labour is not noticed among Rajulu caste group.

Leasing-Out Land

It is important to understand what categories of persons/ individuals lease-out their lands for *paalu vyavasayam*. Generally, the category of people who lease-out agricultural land are widows, employees, villagers presently staying away from the village, and other medium and above category of landholders. Actually, leasing out land is dependent upon many factors. The following few case studies reveal the circumstances under which agricultural lands are leased out for *paalu vyavasayam* in Godilanka.

Rudraraju Savitri is a 58 years old widow and she has two married daughters. She possessed 2 acres of paddy land and 1.4 acres of coconut garden. Earlier her husband used to cultivate agricultural land by employing labour. After his death, she went and lived with her sister at Yedureswaram (about 30 Km from Godilanka). She leased-out the 2 acres of paddy land to Saibaba, a Mala caste person who was earlier working for her family as *sonthamanishi*. Saibaba continued to work in her fields as *paalu* farmer since then. After each crop, Savitri used to visit Godilanka and collect her share in the produce. After a period of five years, Savitri returned to Godilanka. During this period, her husband's elder brother's son used to supervise her fields. She has been giving the hay to Saibaba as she did not possess any cattle. In return, he was looking after the coconut garden. He arranges labour for harvesting coconuts. In addition to this, he helps her in errands like payment of taxes, procuring essential commodities in weekly market, etc.

Rudraraju Ramakrishna Raju, aged 38 years, is a teacher working in Turupulanka, a neighbouring village. His father is 62 years old and now that he became old, the five acres of agricultural land that he owns is not totally cultivated by him. Out of this five acres, three acres were leased-out to two brothers from Vasarlagudem, *Kadali* Purushotham Rao and Yesubabu, from Settibalija caste for un-attached *paalu vyavasayam* by Ramakrishna Raju. The remaining two acres is kept for self-cultivation. Ramakrishna Raju has leased-out agricultural land for *paalu vyavasayam* on the pre-condition that the *Paalu* farmers had to work for him in his field during peak agricultural season (a kind of *sonthamanishi* relationship). During the peak agricultural season not only the un-attached *paalu* farmers have to work for Ramakrishna Raju but they also have to arrange casual labourers. The landowner keeps the paddy straw with him, as the un-attached *paalu* farmers do not possess any cattle.

Alluri Varahala Raju, aged 54 years, owns 11.4 acres of paddy land and 24 acres of coconut garden. He has two sons, both of them are married and employed and live away from the village. Varahala Raju is president of PACS, Godilanka. He leased-out his entire paddy land to four persons. Out of this, 5.4 acres was leased-out for *paalu vyavasayam* to two persons, one to a relative (3 acres) and another to Settibalija (2.4 acres) tenant; and the rest for *gutta vyavasayam* to two Rajulu caste members (3 acres each), who are related to him. Both the *paalu* farmers are treated as *sonthamanishis*. The Rajulu *paalu* farmer supervises the fields and also arranges for labour during coconut harvest days every month. Both the *paalu* farmers attend the work of the landowner during the coconut harvest days. The landowner has bullocks and milch animals. He uses bullocks for ploughing coconut garden and milch animals for his domestic use. As he has more cattle, he collects half of the paddy straw from *paalu* farmers and a quarter of the paddy straw from *gutta* cultivators. He employed a *paleru* from Kapu caste, who comes from the neighbouring village, Godi, to look after as well as milch his cattle.

Thus in *paalu vyavasayam* some changes are noticed with regard to the terms and conditions after the emergence of aquaculture. The changes are basically adaptive changes to the altered situation. Though the above appear to be qualitatively different, largely they retain their earlier character. Apart from this, new type of tenancy, such as *gutta vyavasayam*, has also come into vogue.

Gutta Vyavasayam

Aquaculture in Godilanka has encouraged many progressive and prosperous farmers to venture into this money-spinning enterprise. Those who got into aquaculture lacked time to oversee the agricultural operations. This situation has led to a new type of leasing-out of agricultural land known as *gutta vyavasayam* (contract farming). An important effect of technological progress seems to be leading to the decline of share cropping and a rise in fixed arrangements, like the *gutta* in Godilanka (Swamy, 1988: 565-66; and Shankar, 1999: 3264).

After the introduction of aquaculture in 1993 a considerable amount of agricultural land was converted into aqua farms. The aqua farmer had little time to supervise his fields, as aqua farming requires much time to monitor, in addition to heavy investment. As a result of aquaculture, many a *sonthamanishi* had been displaced. The Rajulu landowners practising aquaculture had accommodated some of the displaced *sonthamanishis* from their caste as *gutta* farmers in their agricultural lands. Under *gutta vyavasayam* the tenant had to bear the cost of agricultural inputs and labour. Apart from this, he has to pay a fixed amount of grain, irrespective of the yield, as lease payment to the landowner. For the first time *gutta vyavasayam* had emerged in Godilanka after the introduction of aquaculture. Under *gutta vyavasayam*, 16-17 bags of paddy per acre annually had to be paid to the landowner (for *dalva*, *sarava* and the third crop) by the *gutta* farmer.

In the succeeding years after 1993, more land was converted into aqua farms. Many marginal Rajulu and Settibalija farmers, who were earlier supervisory *sonthamanishis* to Rajulu landowners and were doing *paalu vyavasayam*, have taken to *Gutta vyavasayam*. Consequently, there was an increase among the tenants who opted for *gutta vyavasayam*. Due to this increasing demand for *gutta vyavasayam*, there was a slight increase in the *gutta* payment per acre annually.

In *gutta vyavasayam*, the tenant is responsible for all aspects of cultivation. The agreed amount of produce (rent) is to be paid after the harvest. The amount of produce agreed upon as rent is negotiated between landowner and the tenant. This payment is dependent on the quality of land, the availability of irrigation facilities and bargaining capacity of the respective partners. It also depends on the relationship between the tenant and the landowner.

Thus there was a growing propensity among the small and marginal farmers and landless tenants from Rajulu and Settibalija castes towards *gutta vyavasayam*. The acceleration of *gutta vyavasayam* was also due to the advancing of money by the commission agents to the *gutta* farmers. At present, there is competition for agricultural land for *gutta vyavasayam*. Switching over to *gutta vyavasayam* liberates one from the bondage of being *paleru* or *sonthamanishi*. Thus, unlike the *paalu* farmers, *gutta* farmers are not obligated to work as labourers in the landowners' fields. Procuring agricultural lands for *gutta vyavasayam* is based more on demand and supply, and impersonal relations. The increasing demand and limited land to be accessed for agriculture has increased the *gutta* payment from 16-17 to 21-24 bags of paddy per acre annually by 1998. The third crop (April-May), pulses and *rotta*, is not shared with the landowner. The third crop determines annual payment of *gutta*, i.e., whether the tenant or the landowner retains the crop.

The sharing of paddy straw depends on the *maatalu* (terms of agreement) between the landowner and the tenant. The tenant, in principle, has a right over the paddy straw. However, in *maatalu*, they might agree either to leave it solely to the tenant or to share it between them. Sharing the straw depends on the ownership of cattle by them. If both the tenant and the landowner have cattle, they share it equally.

Depending on all the above factors the *gutta* payment (rent) fluctuates between minimum to the maximum, i.e., from 21-24 bags of paddy. Lease conditions differ from tenant to tenant, which is more in accordance with their economic conditions. The non-economic factors, such as relationship between landowner and the tenant as *sonthamanishi* for years, also many a time counts in the *gutta vyavasayam* agreement, though indirectly. The agreement is always oral and it is made in the presence of *peddamanushulu* (elderly members) of the contending parties. Any breach of contract will be settled in the presence of these *peddamanushulu*. *Gutta vyavasayam* is practised by both individually as well as jointly.

In addition to individuals, two or more persons lease-in agricultural land jointly from a landowner and practise agriculture on a sharing basis. They pay a fixed amount of grain, or, in some cases, cash equivalent to the grain. The individuals involved in joint *gutta vyavasayam* are mostly relatives, in many a case brothers. They pool-up their resources to venture *gutta vyavasayam*. After paying the *gutta* to the landowner, after the harvest, the partners share the rest of the crop according to their investment of labour and inputs. This sort of farming is observed both among Rajulu and Settibalija caste tenants. In Kharif 1997 101 acres was in *gutta vyavasayam* out of which Settibalija caste persons (60 acres) cultivated a major portion and Rajulu (39 acres) predominate the rest. In both the types of *gutta vyavasayam* 63 acres was cultivated by individuals and the rest under shared. Among Settibalija, *gutta* by individuals was practised in 42 acres and shared in 18 acres whereas for Rajulu it was 21 acres and 18 acres (Table 3.6).

Paalu and *Gutta Vyavasayam*

The category of tenants involved in either *paalu* or *gutta* forms of tenurial arrangement in Godilanka are from different socio-economic backgrounds. *Paalu* or *gutta vyavasayam*- is dependent on the caste and class status of the tenants, the type of land and the source of credit or availability of cash.

The type of land leased-out for *paalu* is generally *saudu polamu* (saline land), a comparatively less fertile agricultural land located near Kousiki, the drainage canal. These lands are in a lower elevation and lack proper drainage. In *saudu polamu* the yields are low and hence, not preferred by the tenants for *gutta vyavasayam*, which is usually carried out in *gandera polamu*. *Gandera polamu* lands are highly fertile, with good drainage facility. Water does not get logged even during heavy rains in these lands. These lands are preferred for *gutta vyavasayam* as the yields are usually high.

It is observed that *paalu* farmers predominate over *gutta* farmers and among them Malas predominate over Settibalija and Rajulu farmers. Among Rajulu only a few are engaged in this type of farming (Table 3.7). The *paalu* farmers are landless and do not possess enough finances for procurement of agricultural inputs.

Presently, most of those earlier *sonthamanishis* in agriculture are now engaged as aqua watchmen. There is a greater demand for agricultural labour, particularly attached labour (*sonthamanishi*), now. The demand for labour also has resulted in an increase in wages.

In addition to getting consumption loan, *sonthamanishis* collect dried coconut leaves for firewood. They are also allowed to collect green grass, which grows on the bunds, for their cattle. Also, they receive gifts from the landowners during the festival days. Hence, many among Mala and Madiga castes prefer to work as *sonthamanishis* and consequently as *paalu* farmers. The reasons for not opting to *gutta vyavasayam* are lack of finances to procure inputs and fear of crop failure, which will make them more indebted, leaving them to the mercy of landowners.

Rajulu and Settibalija caste farmers prefer to lease-in land for *gutta vyavasayam*. Rajulu consider working, as attached labour (*sonthamanishi*) as demeaning. Though Settibalija members worked as *sonthamanishis* for years with landowners, with the experience and understanding of agriculture they have, they now prefer to work as *gutta* farmers.

As the agricultural land is limited, After aquaculture, there was shrinkage of land for agriculture. There is always a competition among various members of Rajulu and Settibalija for obtaining land for *gutta vyavasayam*. And it was the prerogative of the landowner to give his land for gutta. This also depends on the factors like caste hierarchy and economic standing. For instance, if there is a competition for *gutta* between a Rajulu and a Settibalija landless the Rajulu landowner prefers his caste man; and if the competition is between a landless Rajulu and a land-owning Settibalija, the Settibalija is preferred. Thus, in *gutta* the relation is both social and economic.

VII

Agricultural Labour

In a broad sense agricultural labour can be distinguished into familial labour and hired labour. Tradition, customs, caste status, etc., prohibit some of them to work as hired labour on others' farms. The familial labour comprises of those who work only on their own farms. Family labour is not a homogeneous entity.

The agricultural labour work for wages on the lands of landowners and tenants. An essential feature of labour in Indian agriculture is its heterogeneity. The labour employed varies with the kind of cultivation (wet or dry), type of crop (food/ cash crops) and also with the type of production (subsistence or commercial). This heterogeneity is dependent on the variation in the productive qualities of labour (in as far as they are treated as commodities) and the labour market. In this regard, non-economic factors like customs and traditions play a significant role.

In Godilanka, agricultural labour, working for wages, are categorised as casual labour and attached labour, depending on the type of work and the mode of payment. Ashok Rudra (1982) views that a casual labourer is usually a labourer whose duration of contract is for a day and who receives his payment at the end of the day. In Godilanka, the casual labour is locally called as 'private' (independent) workers and they work at their own will with any cultivator. The attached labour is demarcated into *paleru*, *sonthamanishi* and tied labour for a specific crop season. In both the categories of *paleru* and *sonthamanishi*, the recruitment of individuals will be either on hereditary basis or on contract. In the process of time, the later might become hereditary *paleru* or *sonthamanishi*. In case of tied labour, one becomes a tied labour due to some economic reasons, such as borrowing of money, and he might work for one season as compensation for the money borrowed.

For both *paleru* and *sonthamanishi* the duration of contract is normally a year with payment of some occasional payments of cash or grains as a consideration, in addition to the annual payments. He is often called attached labourer as he normally develops certain relations of attachment and dependence on his employer (landowner). If annual contracts of a labour continue for a long period with the same landowner, the tendency of dominance of the employer over the labour becomes a characteristic feature of the relationship.

While dealing with agricultural labour and landowner relations, many scholars commented that caste is an important factor to enter into labour contracts. In Godilanka, it is important to note that Rajulu caste persons will not work under any other caste landowners other than Rajulu landowners. Besides, caste also plays an important role in the way the landowners treat the labour from different castes. For example, a Rajulu *sonthamanishi* may not be assigned any menial job by the Rajulu landowner; instead, he may be treated as a supervisory attached labourer. The agricultural labourers of Godilanka, by and large, belong to Mala and Madiga castes (Table 3.10).

Among the agricultural labourers, more than half is 'private' (independent) labourers and the rest are attached labourers. In case of attached labourers, majority of them are *sonthamanishis*. *Paleru* and *sonthamanishi* are observed more among the lower castes than among the others (Table 3.10 on page 106).

Among the agricultural labourers there exists a hierarchy. 'Private' labourers are considered to be on the top of this hierarchy, followed by *sonthamanishi* and paleru. The category of tied labourers for a crop season fall in between 'private' and *sonthamanishi*. The 'better off' always prefer to work as 'private' labourers, while the others, depending on their economic position, choose to work either as *sonthamanishi* or *paleru*. The details of each category of agricultural labour their duties, payments, work timings their relationship and mutual obligations of the labourers and the landowners are discussed below.

Table 3.10: Agricultural labour* according to caste

Labour Type	Rajulu	Kapu	Golla	Setti balija	Mala	Madiga	Total
Paleru	-	-	1 (6)	2 (12)	8 (50)	5 (31)	16 (100) (10)
Sontha manishi	3 (5)	-	-	10 (18)	38 (68)	5 (9)	56 (100) (36)
Tied	-	-	-	1(4)	23(82)	4(14)	28(100)
Private	2 (2)	2 (2)	4 (5)	12 (15)	53 (64)	10 (12)	83 (100) (54)
Total	5 (3)	2 (1)	5 (3)	24 (16)	99 (64)	20 (13)	155 (100) (100)

* Residents of Godilanka only.
Figures in parenthesis are percentages.

'Private' (independent) Labour

In Godilanka, 'private' labour is independent or free agricultural labour. They have definite work timings, in the morning 8 am to 12.30 p.m. and in the afternoon 2 to 5.30 p.m., a total of 8 hours of work. These 'private' labourers are employed for both agricultural and non-agricultural. The non-agricultural activities include house construction, repairing of thatched roof, clearing irrigation canals, etc. The cultivators require them mostly during peak agricultural operations. During this period there will be more work and the demand for labour will also be more. As a result, wages during these times will be higher than the slack agricultural season. They are also employed during coconut harvest days. Thus, wages for 'private' labour are not constant and predetermined and the wage rate varies between Rs. 40/- to 50/- per day. They are usually paid in cash, but during harvest days they are paid in cash or equivalent in kind (paddy).

Tied Labour for a Crop Season

The number of tied labourers for a crop season increases as one goes down gradually with the category of farmers from big to marginal farmers (Table 3.10). As observed earlier, a few 'private' (independent) labourers work for a landowner for one crop season against a consumption loan received from the landowner. The amount of credit depends on the size of the family of the labourer. The usual loan amount advanced varies between Rs. 500/- to 1000/-. Once he becomes a tied labourer, his family labour also becomes obligated to the landowner. A tied labourer and his family are always underpaid as in case of *sonthamanishi*. For all practical purposes he is treated as a *sonthamanishi*. Tied labour is engaged before the commencement of agricultural season and not during any time of the year. This type of labour is of more recent origin and has been observed only after the introduction of aquaculture in the village.

Sonthamanishi

They are engaged in agricultural operations such as tilling, transplantation, weeding, cutting, and harvesting. In coconut gardens they are employed for collecting, loading, unloading and transportation of coconuts to the landowner's godown. They also receive consumption loans from the landowners, which keeps them tied to landowners. Their work hours are generally from 7 am to 12 noon in the morning and from 2 p.m. to 6 p.m. in the post-noon, a total of about 9 hours of physical work in the fields. The payments are usually in kind (grain). In Godilanka they give 6 '*maanelu*' per day and Rs. 5/- as '*beta*' (a cash payment given daily to meet their individual requirements of tea, snacks, and smoking).

In the category of *sonthamanishi*, two types of arrangements have been observed those who have been working with the same family for generations (hereditary continuation) and those who work on a contract basis. The first type is seen among big, large, and few medium farmers. In the later case, depending on the agricultural work, the landowner hires required number of *sonthamanishis* for a year, which might continue later. Their continuation depends on mutual understanding and also depends on the credit advanced by the landowner. The second type of *sonthamanishi* will be free to switch over to other landowners once the contract is fulfilled.

Every landowner has any one type of *sonthamanishi* who will be at his disposal to coordinate and discharge the agricultural activities on the field. If a landowner has more than two *sonthamanishis* one of them will be regarded as a confidant of the landowner and acts as a supervisor in the absence of the landowner. This is dependent on the length of service of the *sonthamanishi* with the landowner, the trustworthiness and, many a time, caste of the *sonthamanishi* also. He receives wages on par with the others from this category. Besides, depending on the landowner's consideration, he might receive some additional grains after the harvest of the crop as a reward for the faithfulness and for the coordination of the agricultural activities during the agricultural season and coconut harvest.

As part of the contractual arrangement, landowner advances an interest free consumption loan to *sonthamanishi*, which varies between Rs. 2,000/- to 3,000/-. He is given money every week for his family expenses. Though every month his accounts are reviewed, they will be settled only at the end of the year by the landowner. Landowner retains part of *sonthamanishi*'s wages towards recovering credit. Depending on the situation, the contract is renewed by the landowner and it is always oral in nature.

Paleru

In Godilanka, the person who looks after cattle on an annual basis is called '*paleru*'. Though the work of *paleru* is mainly to look after cattle, he also attends domestic work in the landowner's house. His day's work starts at dawn with cleaning the cattle shed and feeding the cattle. He is then given breakfast, invariably leftover rice of the previous day, and tea. Later he sweeps the courtyard, draws water from well or fetches it from the village tank for all household purposes, and cleans utensils. At noon he is given meal. He cleans the cattle shed inbetween, as and when required. In the afternoon, after lunch and after a brief period of rest, he goes to the landowner's fields, cuts green grass and feeds it to the cattle. Around 4 p.m., he washes cattle either at village tank or in the cattle shed. Later, he milches buffaloes/cows and again feeds them. Inbetween, he attends to the other errands of landowner's house. He returns home after dusk. Thus he works from dawn to dusk for the landowner, working for about 12 hours a day. At times, as and when the situation demands, he works for more than the stipulated time.

Paleru is paid either in cash or kind or both on an annual basis. The contract is usually in terms of grain equivalent (bags of paddy). A *paleru* receives 23-25 bags of paddy annually as payment for his services to the landowner. He receives interest free loans from the landowner to meet his consumption as well as other needs, such as ill health of any of his family member, life-cycle rituals, etc., which is usually less than his annual contracted payment (a *paleru* usually receives loan up to Rs. 5,000/- to 6,000/-). *Paleru* thus gets obligated to work with a landowner.

An account for the money advanced to *paleru* will be maintained by the landowner and it will be settled annually as in the case of *sonthamanishi*. After settling of accounts, if the landowner has to pay he will clear them and if the *paleru* has to pay he again renews his contract to work as *paleru* with him for one more year. It has been observed that most of the times a *paleru* spends more money than the wages agreed up on. The contract can be broken and accounts can be settled when the actual work is not in conformity with the agreed terms and conditions. After the termination of the contract the dues will have to be cleared within a month's period. It is usually the *paleru* who gets indebted to the landowner and if he dose not clear the money within the stipulated period he will be charged the prevailing rate of interest in the local area on the loaned amount due from him. In case the landowner defaults, *paleru* can call for a settlement through *peddamanushulu* of landowner's caste. However, such cases are not reported in Godilanka.

As observed earlier, among the various types of attached labourers employed by different farmers, *sonthamanishi*s preponderate over others. Attached labourers from the neighbouring villages are also working under the landowners. Among them, *paleru* are relatively more in number and most of them are from middle ranking castes (Table 3.10).

In addition to the above types of labour observed in the village, a category of organised labourers, *Bantapani* group, from outside started coming to Godilanka after the emergence of aquaculture.

Bantapani

It is a system of collective labour–contract. The mechanization of agriculture has drastically cut down the labour involved in various agricultural operations. The displaced labour that would otherwise be engaged in agriculture have now in this region (within the Mandal) organised into *banta* or contract group labour. They work collectively as a group in the nearby villages during peak agricultural operations such as transplantation, harvest, etc. After the emergence of aquaculture in the region they are engaged in earth works, i.e., conversion of agricultural land to aqua ponds.

Table 3.11: Labour Employed by Farmers during 1997-98*

Farmer type	*Paleru*	*Sonthamanishi*	Tied labour
Marginal	-	-	9
Small	4	12	6
Medium	12	14	5
Large	7	22	3
Big	6	18	5
Total	29(100)	66 (100)	28(100)
Godilanka	16 (55)	56 (85)	28
Outside the village	13 (45)	10 (15)	-
Total	29(100)	66 (100)	28

*This excludes 'private' labour employed by different categories of farmers on different days during the agricultural season.

Banta is a kind of collective piece-rate system in which contract will be between a cultivator, who needs a task to be done faster and does want to maintain attached labourers for maintaining his fields, and the leader of *banta*. Based on the quantum of work *banta* leader estimates the amount of labour required and the amount of money to be paid by the landowner. After this, the landowner bargains and strikes a deal with the *banta* leader. Payment will be shared equally among the members of the *banta* group. The expenses incurred on *banta* group towards tea and other eatables during the contracted period would be deducted before sharing the money. This is a piece rate system and not a time wages mechanism.

There are three *banta* groups engaged in work by farmers of Godilanka. They are from Allavaram (one kilometre from Godilanka) Godithippa (half kilometre from Godilanka) and Komaragiripatnam (two kilometres from Godilanka). The *banta* group is all male and majority of them are from Mala caste. However, there are also a few from other caste groups like Chakali and Settibalija. While working in Godilanka, they employ women labourers from the village.

The *banta* group had first worked in Godilanka during Khariff season in 1996. During this season the agricultural labourers wanted their wages to be increased which were not agreed to by the landowners. Consequently, the labourers refused to work for the existing wages. This had made farmers to bring *banta* groups to work in the agricultural fields for transplantation. This issue of increase in wages was later settled with local political intervention by increasing the wages to a certain extent than to their desirable level. Some of the farmers from then on had preferred to engage *bantapani* group for agricultural activities during peak season.

At first, there was resistance from the labourers in the village. They wanted to pressurise the farmer and persuaded the *banta* group men not to work. They could not coerce them not to work because majority of the *banta* group members belonged to the same caste group. Some of the *banta* group members' even had relatives in Godilanka.

Besides the agricultural operations, *banta* groups were also employed in digging ponds for aquaculture and, more importantly in cleaning the bottom of the pond. This is necessary for an aqua ponds of be cleaned after two or three crops as the bottom would be putrefied with the nitrogenous waste material and the degradation of un-consumed feed which settles on the bottom of the pond. This affects the subsequent yields of aqua crop and so has to be cleaned.

The *banta* group work collectively on a contract basis. The contract specifies the task to be performed and this involves a time limit within which the work has to be finished. The leader of *banta* group representing the workers and the employer negotiates the contract. After reaching an agreement, the work will be initiated. The leader of the *banta*, depending on the work, divides the entire *banta* into small groups comprising 3-4 members and allocates work to the group members. They view that this way of working maintains the tempo of work. The deliverance of services will be efficiently handled with a healthy competition among different small groups.

A group after completion of their allotted work will not help others but rest. Here the works start early at about 7 am and goes till 5pm with a lunch break of one hour. One among them prepares tea once in the morning and also in the evening. In *banta* group they work harder than the casual labour (private) and on an average they earn about Rs. 60/- to Rs 80/- per day and, sometimes, more.

VII

Traditional Payments

The customary payments continue to be still stronger with the attached labourers (*paleru* and *sonthamanishi*). In the earlier days, the landowners after harvest used to leave out some grain in the *Kallam* for the *sonthamanishi* working on their fields. The grain thus left used to be a bag of paddy. Similarly, he used to get such payments at the end of every crop season. The *sonthamanishi* used to receive gifts/money during festive occasions, which used to be spent mostly on toddy. This, now a day, depends on their relationship and the length of time the *sonthamanishi* had been working with the landowner. The longer the relationship the greater is the possibility of receiving traditional modes of payments.

It was customary for the landowner to give grain (part or total) on the occasion of life-cycle ceremonies, such as marriage or death in the family of *paleru* (hereditary or otherwise) working with the landowner for a longer period of time, say about 10 years and above. It also depends upon the number of persons working under a landowner from the same family as *paleru*. If both father and son work as *paleru* under the same landowner, it is obligatory for the landowner to provide the *paleru* family grain for the occasion. In case of *sonthamanishi* this kind of payments are not common.

All the attached labourers (*paleru* and *sonthamanishi*) irrespective of hereditary or contracted, will receive two pairs of clothes annually. One pair is given during Sankranti, which is a major festival in the area, and the other either during Dasara or Deepavali festival. The attached labourers are given a locally stitched banian, lungi (cloth worn by men below the waist) and *kanduva* (towel), which is tied to the head while working. In addition to these customary payments, the attached labourers receive monetary payments as gift during festivals and fairs. Apart from these, the attached labourers receive special food prepared during festivals in the landowners' houses.

At present, many customary payments are translated into monetary terms. A *sonthamanishi* does not get anymore grain left in the *Kallam*.

Most of the attached labourers are unwilling to work under the same landowner with whom their fathers had worked. There is also some change in the modes of payments by the landowners. The attached labourers working for short duration do not receive any payments during lifecycle crises. However, money advanced during such crises is accounted for as credit. Thus an impersonal and contractual tendency is increasing with regard to the relationship between the landowners and the attached labour.

Vadrangi (Carpenter-cum-blacksmith)

He is a resident of Penumalapeta in Godilanka and acts both as carpenter as well as blacksmith. He repairs bullock carts, ploughs, and other agricultural implements of the farmers. He is paid a bag of paddy for a *jata* (pair), implying a bullock cart, wooden plough, and small implements used in agriculture and coconut garden by a landowner. The dependence of farmers of Godilanka on wooden plough is now declining due to mechanisation of agriculture and this is further accentuated by aquaculture. However, wooden plough is still used for ploughing coconut gardens. Marginal and small farmers continue to use wooden plough in their agriculture. The role of *vadrangi* has declined significantly when compared to the situation prior to 1980s. Many farmers now prefer payment not on *jata* basis but on piece-rate and in cash based on the number of agricultural implements serviced by the *vadrangi*. The piece-rate payment is found to be beneficial to medium, small or marginal farmers, as they will not be tied up to the *vadrangi*, in a customary sense, on an annual basis. However, the large and big farmers still continue to maintain the traditional and customary relationships. There is an increase in grain payment to the *vadrangi* for one *jata* of implements. At present, for a *jata* of agricultural implements one and half bags of paddy are paid annually.

Mangali (Barber)

Earlier, *Mangali* used to visit the houses and render his services, *kshavaram* (haircut and shave). A household used to give him an annual payment in kind (grain). Depending upon the number of persons to be served in a household, the grain payments were worked out. However, a household used to give him a minimum of half a bag of paddy for *kshavaram*.

At present, Mangali still continue their traditional services for the older generation in the village and the payment is made still in kind annually. The youth and middle-aged persons, who went out for education and other purposes, have learnt shaving their beard. Mangali visits the houses of such persons for haircut and he receives payment on piece-rate basis and in cash, but not in kind. The first haircut saloon was opened in this village near Gogulamma (village deity) temple. Recently (about five years ago) in Vasarlagudem two saloons were opened. At the saloon the relationship becomes impersonal and service will be rendered on the basis of payment in cash.

Chakali (Washer man)

Godilanka, since its inception, does not have a *Chakali* residing in the village. The *Chakali* from neighbouring villages (Godi and Mogallamuru) render their services to Godilanka. Earlier, the payments were in kind. For an average household of 5-6 members, clothes were washed once a week and payment was 2 bags of paddy annually.

At present, the payment is calculated for every 100 items washed without any distinction of size and quality of the cloth and the payment varies from Rs. 80/- to 100/-. The swiftness of the service is dependent upon the swiftness of payment of cash by the client. Some households from large and big farmers still follow the old system of payment in kind. The *Chakali* prefers his annual payment in cash equivalent terms.

All the servicing castes, such as *Mangali* and *Chakali*, receive food and gifts in cash during the festivals. *Chakali* and *Mangali* have customary rights and roles to play during life-cycle crises of Rajulu, for which they receive special payments, stipulated amount of grain or cash, and gifts.

VIII

It is important to study the sources of credit in a village. The relationships in the village are defined and sometimes redefined based on the sources of credit and its availability. Economists conducted many studies on informal credit in Indian agriculture. They had laid emphasis on issues like demand-supply determinants of interest or on the extent and nature of indebtedness. Some of the studies attempted to relate the extra-economic factors to informal credit, as they had a bearing on the economic variables. The 'forced commercialisation' and 'distress sale' in the backward regions of agriculture are a result of the nature of credit, which, in reality, is "based on a system of relations that arc cssentially social" (Bhaduri, 1984: 51). A similar view was held by Bharadwaj (1982: 27), who pointed out that 'the relative bargaining strength' depends, also gets reinforced, on the traditions, customs, and social norms which determine the access to resources, the conditions of receiving, etc., in an agrarian economy.

Many Anthropologists and Sociologists had referred credit and their relation to the social process in their studies on patron-client relationships (Srinivas, 1955; Epstein, 1962); village economy (Béteille, 1969); factional politics (Lewis, 1958); production relations (Gough, 1955, 1960; Marriott, 1960; Breman, 1974; and Brass, 1986); and, peasant discontent and mobilisation (Moore, 1966; Desai, 1979; and Dhanagare, 1985). Due importance was not given to the cultural or societal view to deal with informal credit or debt relations. The needs of villagers, their credit relationships, the nature and extent of indebtedness are important to be understood for a better understanding of agrarian relations. The credit relations mediate the existing relations among various categories of people involved in agriculture.

Village Credit Structure

The rural credit that exists in Godilanka comprises of both institutional and non-institutional credit. They are otherwise referred to as formal and non-formal credit. The institutional sources of credit include PACCSs, commercial banks, etc. The non-institutional sources of credit available to the villagers are moneylenders, *sahukars*, commission agents, merchants, relatives/ friends, etc. The institutional credit organisations provide crop, gold and land mortgage loans. There exists a uniform rate of interest to all borrowers of either short-term or long-term loan needed for agricultural purposes. Short-term loans are required for meeting the crop season requirements, where as long-term loans are required particularly for investment in farm machinery. Loans taken under the IRDP Scheme sponsored by the Government has a subsidy component, which varies according to the caste background of the recipient (33% for Scheduled Castes and 25% for BCs). In Godilanka, majority of the people avail credit from non-institutional sources and thus, our study deals with the non-institutional/informal credit.

In the non-institutional credit the interest is not fixed but varies according to the loan amount and the borrower. The major providers of informal credit in the village are moneylenders and merchants. Besides, individual households also borrow from either relatives or close kin in the village or outside the village. A large number of loans advanced in the village, including informal credit, are tied to some contract or the other.

As noticed earlier, it is also common that landowners give credit to villagers. The rates of interest for the loans (informal) borrowed from a single moneylender/landowner are not uniform and in certain cases interest is not charged. This type of consumption loan without any interest is linked or tied to contracting agricultural labour. In other words, consumption loan is used as bait to restrict the agricultural labourer to work with the landowner.

The credit relations many times are mediated through the existing relations among the various categories of people working in agriculture. These relations to a large extent are between agricultural labourers and landowners and between landowner and sharecroppers.

For a cultivator the need for credit is two folded. He may need it for production purpose as well as to meet the consumption requirements of his family. In case of agricultural labourers, they need loans mainly for consumption purposes, which include spending on life cycle events, illness of a family member, compulsion to help a relative(s), etc.

Getting informal credit in Godilanka was observed to be inversely proportional to the land-ownership. Most of the marginal farmers get their credit needs for agriculture from commission agents, whereas large farmers get it from *sahukars*. In case of medium farmers, they obtain it from both *sahukars* as well as moneylenders (Table 3.12).

The *sahukar* in Godilanka is mostly associated with a business, provision store or kirana shop. He mostly advances loans to small and marginal farmers. He supplies credit to farmers against property either in the form of agricultural land/ coconut garden or a residential house. Just before commencement of the season, farmers approach the moneylender for loan, which is essential for the crop season. The amount of loan depends on the money the landowner possesses and the extent of land that he would be farming. If the *sahukar* does not have sufficient money, the farmer then approaches another *sahukar*. The loan will be advanced at different stages of the crop by the *sahukar*. After it is agreed upon to advance the loan, it is entered in a notebook, which contain the details of loan and the rate of interest, which varies from Rs. 3/- to 4/- per 100 per month, depending on the size of loan. This notebook will be an evidence for the moneylender, if a conflict/non-compliance of conditions by the loan arises. The farmer has to repay the loan money after the harvest of the crop. The *sahukar* expects repayment of credit with interest in terms of cash, at times he even accepts it in kind.

Table 3.12: Informal Credit Advanced to Farmers during 1997-98

Type of farmer	Commission agent	Sahukar	Moneylender	Total
Marginal	20(91) (77)	2(9) (8)	-	22(100) (39.3)
Small	5(27.8) (19)	12(66.6) (50)	1(5.6) (17)	18(100) (32.1)
Medium	1(8.3) (4)	8(66.7) (34)	3(25) (50)	12(100) (21.4)
Large	-	2(50) (8)	2(50) (33)	4(100) (7.2)
Big	-	-	-	-
Total	26(46.4) (100)	24(42.9) (100)	6(10.7) (100)	56(100) (100)

Figures in parentheses are percentages.

There are 8 *sahukars* in Godilanka and during 1997-98, for both the Rabi and Kharif seasons, they advanced loans of Rs. 98,600/- for small and marginal farmers. In addition to agricultural loan, they advanced credit also for non-agricultural purposes.

Commission Agents also supply credit to cultivators, but the credit is advanced for agricultural purposes only. The process of loaning is similar to that of the *sahukar*. The commission agent supplies credit to medium, small and marginal farmers as well as tenants on a pre-condition that the harvest reaped shall be marketed through him. This will be a security for him. The usual rate of interest is more than that charged by the *sahukar* and varies from Rs. 5/- to 6/- per hundred per month. After the harvest he procures the entire paddy from the farmer's fields and transports it either to a rice mill or to another commission agent of bigger size. Later, the loan amount along with interest will be deducted and the remaining money will be returned to the cultivator. There are 5 commission agents in Godilanka and they provided Rs. 1.23 lakh during 1997-98 for both Kharif and Rabi seasons.

In addition to *sahukar* and commission agent, professional moneylenders are also there in Godilanka, who extend credit to the villagers. There are 6 professional moneylenders in Godilanka and all of them belong to Rajulu caste. These moneylenders provide credit only against some security. The security can be in the form of agricultural land/residential house or ornaments. A document will be executed with the details of credit and the agreed rate of interest. The rate of interest depends on the amount of money required. The rate of interest is usually lower for large borrowings. For instance, if the credited amount is less than Rs. 30,000/- the rate of interest is Rs. 5/- and if it is more than one lakh rupees the rate of interest is Rs. 2/- per 100 per month. The rate of interest fluctuates for the amount in between Rs. 30,000 and one lakh. Unlike the loans from *sahukar* and commission agent, there is no time limit to clear the loan.

In Godilanka credit is given to attached labourers. Amount of credit advanced increases gradually from tied labour to *paleru.* In the category of tied labour who work for a season gets less than *sonthamanishi* who works on an annual basis. Between the *paleru* and *sonthamanishi*, *paleru* receives a higher amount of loan than *sonthamanishi. Paleru* receives a higher amount of loan because the nature of work is tougher than *sonthamanishi* and in order to attract people for this work a higher amount of loan is advanced. In some cases, a much higher amount of loan is advanced to some *palerus* than is usually given to others in that category in order to retain them as they were reliable and were working with the landowner for a longer period of time (Table 3.13).

Landowners, shopkeepers, petty moneylenders, relatives and friends provide credit to the landless agricultural labour. The reason for which agricultural labour needs credit is to meet their domestic and family expenses. This may be for marriage, house renovation, to meet hospital expenses of their family members, maternity expenses, life cycle rituals and other expenses.

Table: 3.13: Type of Agricultural Labour and the Loan Advanced

Type of Agricultural Labour	500-2000	2000-3000	3000-5000	More than 5000	Total
Paleru	-	-	20 (60) (45)	9 (31) (100)	29(100) (23.6)
Sonthamanishi	-	42 (64) (86)	24 (36) (55)	-	66 (100) (53.7)
Tied labour	21 (75) (100)	7 (25) (14)	-	-	28(100) (22.7)
Total	21 (17) (100)	49 (40) (100)	44 (36) (100)	9 (7) (100)	123(100) (100)

Figures in the parentheses are percentages.

'Private' agricultural labourers also approach shopkeepers and petty moneylenders. Petty moneylenders generally lend money in small amounts ranging from Rs. 100/- to 5,000/-. The money is lent against pledging of articles like bicycle, watch, and household articles. They do not execute any document but maintain records, which becomes important at the time of realisation. There are 15 petty moneylenders from different castes, 6 among Rajulu, 3 among Settibalija, and 6 among Mala. The rate of interest they charge is Rs. 5 for hundred every month.

IX

Role of Women in Agriculture

Women are employed mostly in casual agricultural work and they are employed in operations such as transplantation, weeding, and harvesting. It is widely observed that wages paid to women are always lower, quite often substantially, than that paid to men. In the wet rice cultivation areas of South India, a large part of work force of women is seen as hired labourers (Mencher, 1982). Participation of women in work force depends upon the caste/class status of a woman (Béteille, 1971; Bardhan, 1981; and Duvvury, 1989).

In agriculture women work for more number of days than men do. The tasks that women are engaged will be more arduous than those undertaken by men (Table 3.14).

Table 3.14: Sex-wise Distribution of Working Days Per Acre

Operations	Work days	
	Women	Men
Field preparation	-	2
Seed-bed preparation	1	3
Transplantation	10	4
Fertiliser (twice)	-	2
Weeding (2 times)	4	-
Pesticides	-	1
Harvesting	9	2
Threshing	-	6
Winnowing and Transport	2	2
Total	26	22

In Godilanka, women employed in agriculture are mainly from Mala and Madiga castes. However, a small proportion of women from Settibalija caste also works as agricultural labourers. Most of the Settibalija women work only on their family farms. During peak agriculture season women are paid Rs. 30/- per day and in lean season they are paid Rs. 25/-, if employed. It is essential to note, as observed earlier, that if a person is working as attached or tied labourer it is obligatory for his wife and other members of the family to participate along with him in the agricultural work during peak agricultural season in the landowner's lands. For this, they receive Rs. 20/- per day, which is less than the prevailing wage labour in the village.

Presently, most women do not find employment in agriculture because of the shrinkage of land under agriculture, due to the introduction of aquaculture. These displaced labourers work as house and servant maids in Rajulu caste houses.

X

Out-migration

In the later part of sixties and early seventies with the implementation of Tenancy Act and Land Reforms Act many landowners evicted the tenants from their lands. This has reduced the status of the intermediary caste groups, who were earlier working as proxy landowners, to agricultural labourers alongside of lower castes over whom they had authority earlier. Apart from this, the mechanisation of agriculture has also brought down the use of agricultural labour to a large extent.

The above factors forced many tenants to migrate to distant places in search of livelihood. Thus, some members from *Settibalija* caste group had migrated to Kandla in Gujarat as casual labourers because of some relatives in Kandla. While the Settibalija migrated to Gujarat, Malas migrated to different industrial and urban places. To begin with, only men migrated, and, later, after some of them who felt secure took their families also. Due to meagre wages in the place of migration others had left their families behind in the village and used to visit their families occasionally. Most of them used to send remittances, which were invested in buying agricultural land and building houses.

Due to improvement in educational qualifications and other entrepreneurial qualities, many landowners moved to nearby towns and cities either for education of their children or for business. The mechanisation of agriculture had further accentuated out-migration. Mechanisation did not require much supervision of agricultural activity. Most of the educated youth were not inclined to do the arduous agriculture but preferred white-collared jobs in the government sector as well as in private companies. Thus, many youth from the land-owning castes, especially *Rajulu* migrated from the village to nearby towns and other urban centres in the State.

CHAPTER 4

Production Relations in Aquaculture

This chapter deals with a brief sketch of the genesis of aquaculture and its growth in Andhra Pradesh. It also discusses the conditions that led to the adoption and expansion of aquaculture (prawn) by farmers in Godilanka. It attempts to analyse the relations of production in aquaculture and the patterns of land-labour relations after the emergence of aquaculture. It further analyses the role and status of women in aquaculture farming, the relationship between the aqua owner and labourer, credit relations and the new occupations that have evolved after aquaculture in the village. Apart from the above, the impact of aquaculture on the surrounding environment is also discussed in this chapter.

I

Genesis and Growth

Harvesting of aquatic organisms and its products is done in two ways, one by capture, where they are harvested from the natural water (sea, river, ocean, lakes, etc.) using nets, boats, etc., and the other by culture (rearing) in the specially created ponds. In the earlier times, shrimp and its species were harvested by capture using small-mechanised boats from inshore waters and trawlers to exploit resources in the deep sea.

Aquaculture is a system of cultivation of aquatic organisms (fish, crabs, shrimp/prawn, etc.) under artificial environment in small, medium or larger ponds specially designed and constructed for this purpose.

Historical Overview

Aquaculture in Asia has a history of more than 2,500 years. Aquaculture, primarily fish culture, presumably began about 500 BC in China, for catering the needs of the emperor who wanted a constant supply of his favourite fish. The earliest treatise titled Fan Li, dealing with the raising of fish, contained procedures of pond design, ecological considerations, selection of best species, size of mating fish, mating seasons, etc., (Liao, 1992).

It was in the 15th century that brackish water aquaculture began in Indonesia (Liao, 1992) with the culture of Milkfish (*Chanos chanos*). It was observed that all along the coast backwaters (brackish) aquaculture has become a regular part of the rural economy in Asia (Csavas, 1994:8-24).

The technological success of Japan and Taiwan over the past 20 years has served as a model for the Asian countries in the development of aquaculture (Liao, 1992). The establishment of the technique developed in Taiwan of larval rearing of grass prawn (*Penaeus monodon*) in 1968 provided a good basis for culture of prawn species.

Asian nations account for over 80 per cent of the total farmed Shrimp harvests in the world (US, Embassy, 1991; and FAO, 1989). The average growth rate of aquaculture production in Asian countries during 1977-87 was 15% (INFOFISH International 1990). Many Asian countries have a suitable climate and large areas are available for both marine and freshwater Shrimp culture. There are several indigenous species that are cultured in many nations of this region. However, three major Shrimp groups commonly cultured in Asia are marine species belonging to the genera *Penaeus* and *Metapenaeus*, and fresh water species of the genus *Macrobrachium*. *Penaeus monodon*, known as 'tiger prawn', dominates the industry in Southeast Asia.

From times immemorial, traditional methods of brackish water aquaculture existed in India in the States of West Bengal, Kerala, Goa, and Karnataka (Alagarswami, 1991:222 and Muralidharan and Shankar, 1995). In these areas paddy-cum-aqua farming is done in about 50,000 hectares (Sundararaj *et al*, 1992). In the low-lying coastlands in these States one crop of paddy is/was raised during rainy season (low saline period) and during summer a crop of prawn is/was cultivated (Guimaraes, 1989).

Development of Prawn Culture

In the last part of 1980s the prawn production through capture fisheries could not meet the demand. Coastal aquaculture was found to be an alternative and promising area of development a decade ago for increasing prawn production to meet the demand, which is generally export oriented. Prawn culture utilises derelict water for better aquatic harvest providing of productive and remunerative employment to the labour, improving the incomes of farmers (small and medium) and an overall development of the rural economy. The traditional fishermen utilised these resources to eke out their livelihood but this was often seen to be fortuitous. However, little attention was paid for the improvement and development of aquaculture in brackish waters till recently.

Although the Shrimp culture industry in Asian countries boomed through the early and mid-80s, its success was there only for a short period. In the last part of 80s a few among the big suppliers of frozen prawn from South East Asia to the International market could not maintain their place in the export production. In 1988, in countries like Taiwan, diseases spread rapidly, which affected the Shrimp production. This led to crop failure in the succeeding year. In the following year, larvae died a few weeks after being placed in grow-out pond. The harvest thus decreased nearly by half and could not recover even in the next year (US, Department of Commerce, 1981-90). As in Taiwan before 1990, decrease in Thailand central province has caused the production stagnant. Here the prices have doubled at this juncture (US Embassy and Consulate Reports, 1989-1992). The ash falls and mudflows caused by eruption of Mt. Pinatubo in 1991 damaged the grow-out ponds in Philippines (US Embassy and National Marine Fisheries Science, 1990).

All the above aspects led to the decline in production in cultured Shrimp and its demand in the international markets has given an opportunity for the big international feed companies to open their markets in other favourable Asian countries so as to meet the market demand. In this context, the economic Liberalisation and Globalisation had hastened up this process, which led to a phenomenal growth of shrimp aquaculture in India. The Indian Government actively supported the development of shrimp aquaculture in the name of 'blue revolution'. A large number of private companies, transnational corporations, rich landlords and industrialists had invested in Shrimp farms, feed processing plants and hatcheries in a number of coastal villages, especially along the coastal belts in Andhra Pradesh, Tamil Nadu and Pondicherry (Mukul, 1994: 3075-78).

The rising prices in the international markets and growing demand for the Shrimp, besides the declining trend of prawn landings and also hike in cost of fuel of mechanised boat and trawlers, necessitated the inland culture of Shrimp. The availability of species that have commercial value has also attracted the entrepreneurs to develop Shrimp farming by utilising brackish water. Thus, shrimp farming has gained momentum in India.

Trends

India's cultured Shrimp harvests increased from an estimated 15,000 tonnes in 1982 to about 32,000 tonnes in 1990 (Schneider, 1991; FAO Yearbook 1991). The Indian prawn production is around 0.2 million tonnes, which is about 9% of the world's production. Indian farmed prawn (0.025 million tonnes) constituted about 14.2% of the global aquaculture prawn production, which is 0.59 million tonnes (Infofish, 1990; MPEDA, 1989; Coche, 1983).

The non-penaeid prawn prior to 1960s has dominated the Indian prawn capture fisheries. During this period the penaeid prawn production had contributed about 5% and non-penaeid 7.4% to the total marine fish landings. Between 1961-1970 penaeid prawn dominated the total Prawn eaters and attained a maximum of 0.13 million tonnes in 1970. The growth of Indian farmed-prawn was around 115% over 5-years, during 1984-85 to 1988-89, with an average annual growth of 23% in contrast to the global increase of 200% in the same period. In 1989, India stood seventh in the global prawn culture production (Sundararaj *et al.*, 1992).

Aquaculture is justified to provide food for the ever-increasing population at the global level. The products of aquaculture are quite expensive and can be consumed mainly by the middle class and the wealthy from the industrialised countries (De Walt *et al.*, 1993:1204).

Export

There are about 30 fishery products exported from India, shrimp is by far the principal item which is accounted for 60% in volume and 80% in value of all marine products. Indian exports of shrimp during the last decade have varied marginally in terms of volume. The exports have increased from 47,000 to 55,000 tonnes but the value of this has shown a consistent increase from US $ 140 million in 1980 to $ 330 million in 1988 (MPDEA, 1989). The export of Indian marine products increased by 19.4% by quantity and 6% by value between 1988-89 and 1989-90, reflecting the potential that prawn farming holds for the country's economic growth.

The Government of India has realised that there is a vast potential for earning exchange through cultured shrimp exports. Consequently, a national policy was formulated to promote the shrimp culture as an industry in order to retain India's position as one of the leading producer of world shrimp. The Marine Products Development Agency (MPDEA) was established in September 1972 under an Act of Parliament to provide technical and financial assistance programmes that began under the Seventh Five-Year Plan.

MPDEA

The organisation is headed by a senior civil servant of the Government of India, who assumes the position of MPDEA Chairman. MPDEA has field offices established for the promotion of shrimp farming to help farmers in selecting sites, prepare project reports for arranging credit, construct farms, prepare ponds, secure post-larval forms, select stock and manage ponds. Since its creation, the MPDEA has contributed to the growth and modernisation of the Indian seafood industry. It also organises short-term training programmes, farmers meetings and workshops. The MPDEA is responsible for the overall development of the Indian seafood industry with special emphasis on export and production promotion. It has both regulatory and promotional functions.

Types of Shrimp Farming

The process of rearing the fingerlings of shrimp up to a marketable size in enclosed water body or specially created aquatic environment is termed as shrimp culture/farming. In view of the nature of inputs and principles of scientific management, shrimp farming systems can broadly be classified as traditional, extensive, semi-intensive, and intensive types (Muralidharan and Shankar, 1995; MPDEA, 1991).

Traditional

In the traditional system of shrimp farming, shrimp seeds, along with other aquatic forms, are trapped. This happens during high tide when the ponds get inundated. In this process, along with shrimp many competitors and predators get in along with prawn/shrimp and co-exist with them in the same pond. Indiscriminate stocking of both desirable and undesirable varieties of shrimps, fish, etc., lead to unpredictable production and it is often very low in quantity. By fixing suitable screen(s) in the sluice, the trapped organisms are prevented from escaping and the crop is harvested at frequent intervals. Here the shrimps are often harvested before they grow to optimum size. The average annual production from traditional system is below 0.5 tonnes per hectare.

Extensive

The extensive system of shrimp farming is an improvisation of traditional farming. This involves construction of new ponds between one to five hectares in suitable areas, which depends on the availability of seawater. The pond is selectively stocked with fast growing shrimp seed at a comparatively low density and they range from few thousands to one lakh per hectare. The stocked shrimp is given a supplementary feeding and water quality is maintained. The water quality maintained may be from either natural through rise and fall of tides or artificially changing water usually up to 10 per cent by deploying engines. The average production is between one to one and a half tonnes per hectare in a crop season.

Semi-Intensive

In the semi-intensive system, the ponds are artificially constructed with a size from 0.2 to 1 hectare. The ponds are selectively stocked with hatchery seed in relatively higher-density, i.e., from one to three lakhs per hectare. Because of high-density, there is a need to maintain the water quality. About 10-12 per cent of the water is changed everyday. The pond is aerated either with air blowers or paddle wheels. The shrimp is fed with a graded feed at different levels of growth, which is balanced with the requirements for its growth. The average production ranges from four to five tonnes per hectare in a crop season.

Intensive

The intensive system of shrimp farming involves construction of ponds (sometimes concrete), usually between 0.03 to 0.1 hectare in size. Here care is to be taken at every stage of culture. The stocking is selective with quality seed procured from hatcheries, under required conditions, with seed density between five to 10 lakh per hectare. The most important and vital part is maintenance of water quality. About one third of water in these ponds are changed daily with pump-sets and periodically aerated with mechanical aerators for 3-4 hours in a day. The stocked shrimp seed is fed with a nutritionally well-balanced high energy feed. The average production from this type of system ranges from 10-12 tonnes per hectare in a crop season.

Of the above types of shrimp farming discussed, traditional fishermen communities along the coast engage in traditional method. While others methods such as extensive, semi-intensive and intensive are followed by entrepreneurial individuals either in agricultural or in other lands. Mainly farmers and fisherman of small economic potential are observed in extensive type of farming. The remaining both semi-intensive and intensive are carried out by big farmers/corporate companies which involve huge capital and technology in addition to labour.

II

Andhra Pradesh has a coastline of 974 kilometres and is the second largest maritime States in Indian union. Rivers like Nagavali, Vamsadhara, Godavari, Krishna and Penna join the Bay of Bengal providing extensive estuaries and back water resources. All along the coastline there are numerous bays, low lying areas, estuaries, backwaters, brackish water lakes and mangrove swamps.

According to the National Commission of Agriculture, the brackish water areas, including swamps, available in Andhra Pradesh are 0.2 million hectares. Vast brackish water offers a natural resource for the exploitation of fishery, which is influenced by the ebb and tide.

The survey conducted by the State indicated that there are 172 brackish waters with a total spread of 63,962.34 hectares covering 9 coastal districts of Andhra Pradesh. The total area estimated, suitable for brackish water culture, is 17,000 hectares, excluding deep portions of swamps, creeks, drainage courses, thick mangrove forests and sand dunes, considered potential for developing and construction of brackish water farms. These areas belong to revenue, salt and forest departments, and private individuals (Girglani, 1987).

There are new areas that are developed into prawn farms in the states of Andhra Pradesh, Orissa, Tamil Nadu, and Kerala. The development of shrimp farms in Andhra Pradesh can be grouped broadly into three groups. The First group belongs to the middle level entrepreneurial farmers who had developed shrimp farms on the initiative of MPEDA in mid-1980s. The Second were the small or medium farmers who developed the farms after recognising the success of middle level entrepreneurial farmers. The farming done by these persons was not on systematic lines and without meeting basic requirements and also at places on encroached or illegally transacted lands. The Third group is of the big entrepreneurs who started aquaculture on a large scale. They had purchased large tracts of land at high prices and in some cases encroached, illegally transacted or even purchased by coercion. This, however, includes the government allotted lands on lease basis to either individual or cooperative societies (Muralidharan and Shankar, 1995). In quite a few cases in this category, there were integrated projects with hatchery, feed mill and processing mill, in addition to the main farm.

Table 4.1: Extent of Brackish Water Area by State

STATE	Estimated brackish water area (hect.)	1990-91	1991-92	1992-93
West Bengal	405000	33815	33918	334050
Orissa	31600	7075	7417	7760
Andhra Pradesh	150000	6000	8100	9500
Tamil Nadu	556000	250	480	530
Pondicherry	800	Negligible	Negligible	Negligible
Kerala	65000	13000	13145	13400
Karnataka	8000	2500	2500	2570
Goa	18500	525	525	550
Maharashtra	80000	1800	1869	1980
Gujrat	376000	125	231	360
Total	1190900	65100	68227	70700

Source: MPDEA.

The first group though did not farm on scientific lines, it did not cause any problem to others. The second group are almost wayward, follow their own intuitive and common sense knowledge than scientific reasoning. They have problems in the operations pertaining to culture as well as those relating to social and legal arenas. These people without any intention do lot of harm to the ecosystem. The third group grew rapidly with the aid of foreign technology and inputs. They get highest yield possible per unit area. Their operations are on scientific lines only in management of farm as an exclusive unit without any concern for natural environment and with little connection about local people. They operate more or less, on the lines of an industry with skilled personnel recruited from outside and a few locals, who seldom are recruited as watchmen and/or unskilled labour. Godilanka aqua farmers come under the second group.

Table 4.2: Land under Various Areas in Aquaculture

District	Ownership-wise Area (in Hectares)						
	Fisheries	Revenue	Forest	Salt	Port	Private	Total
Srikakulam	346.3	845.2	-	1029.3	200	1502.6	3932.4
Vizianagaram	-	2.5	-		-	9.5	12.0
Visakha	-	414.3	-	86.7	-	490.7	1291.7
E. Godavari	-	4273.5	9532.8	-	96	935.1	4837.4
W. Godavari	-	1583	-	-	-	1884.5	3467.5
Krishna	-	10418.6	8000	-	-	243.4	24838.0
Guntur	-	254.2	1310	-	-	30.6	1594.8
Prakasam	-	1058.5	-	242	-	111	1411.5
Nellore	-	2690.2	-	26	-	36.8	2753.0

Source: State Government Survey, 1982-83

Awareness of Prawn Aquaculture

Prawn culture was observed in Andhra Pradesh in the mid-1980s. This is observed mostly in the coastal brackish water areas found to be suitable for prawn aquaculture. Prior to 1991 the culture of prawn and its species was mostly restricted to salt pans, un-cultivable lands nearer to sea and swamps, which are often affected by the tidal movements either by the intrusion of salt water through canals or effected by earlier cyclones, etc. These farmers depend mostly on the traditional methods of farming and feed rice bran, snails and, at times, beef, etc., as supplementary feeding.

The introduction of the New Economic Policy (1991) and the liberalisation of Indian economy have made import of capital goods easier. As a part of this policy, the technical know-how has been transferred. In addition to this, knowledge and use of feed (supplementary nutrition) has also been imported from foreign food companies, which were already in operation in many Southeast Asian nations. The large industrial firms and corporate companies, who had interest in investing their capital in this lucrative business, to a large extent, supported this. Thus the above conditions have increased the pace of prawn farming to the status of an industry. This lucrative vocation, where profit margins are very high in a short span of time, has drawn the interest of many industrial and corporate bodies, as well as large entrepreneurial farmers (Rao, 1995).

Our study, however, deals only with the farmers involved in the prawn aquaculture in a village of East Godavari district of Andhra Pradesh. Among the districts where aquaculture has developed rapidly after 1991 in Andhra Pradesh, East Godavari district occupies a prominent place in the area under aquaculture and the selected village is one of the progressive aquaculture areas in this district.

Many of the feed-supplying companies encouraged farmers who were inclined to undertake aquaculture. These companies provided technical know-how for the export oriented prawn culture. They entered into a buy-back policy[1]. This policy has attracted most of the land-owning farmers who were progressive and had inclination to undertake aquaculture in the region. The material benefits associated with this new production strategy had a profound impact on many entrepreneurial farmers. With the demonstration affect, many medium and small farmers have also taken to it in a short span of time. Frequent failure of agricultural crops due to vagaries of nature and problems involved in the maintenance of agricultural labour have hastened the process of adoption of prawn farming in this area.

Buy-back Policy

At first, the foreign feed companies and a few Indian corporate companies supplying feed introduced the buy-back policy in many areas of Andhra Pradesh, including East Godavari district. According to this policy (1991-92), any farmer inclined to do prawn-culture has to dig a pond according to the technical specification recommended by the feed company. Besides, experts from these companies also assisted the farmers. Basing on the recommendations of the experts, a contract was signed by the farmers and the company supplied the feed on a pre-condition that the harvest has to be sold to their company only. Hence, it was called as the 'buy-back' policy. The feed company intermittently used to send their technical experts to inspect the farms.

In the beginning, technical experts from Taiwan and Thailand provided initial guidance in raising a crop. In the initial years there was a bounty of harvest and as per the agreement it was sold to the feed companies. This has benefited not only the farmers but also the feed companies to a large extent. Looking at the prosperous aqua farms many others in the region started venturing into aqua farming as the buy-back policy was still in existence. In 1994 there was a heavy loss to aqua farmers due to the effect of virus to prawn crop all along the coast of Andhra Pradesh, East Godavari was no exception to this. Under these circumstances, the aqua farmers could not pay the costs incurred towards the feed supplied to them by the companies. This made the feed companies to sue the farmers, who had signed the contract of buy-back policy, which later was proved that the agreements had no legal sanctity. The feed supplying companies had no collateral security from the aqua farmers who signed contract with them. Under these circumstances, the feed companies withdrew the buy-back policy.

Prawn Aquaculture in Godilanka

In East Godavari district, all along the coast and also in areas where they have accessibility to back waters, most of the landowners converted their agricultural lands into aqua ponds. In the summer of 1993 many progressive and enterprising farmers in Godilanka converted highly productive agricultural lands into aqua ponds. The growth and evolution of aquaculture in Godilanka can be distinguished into four stages. During the first and second stages big landowners were attracted to aquaculture and the later two stages attracted the others (Table 4.3).

Table 4.3: Adoption of Aquaculture in Godilanka

Stages	Year	Farmer type	Number	Acres
First adopters	1993	Big and Large	18	52
Second adopters	1994-95	Large and Medium	26	67
Later adopters	1996-97	Medium and Small	18	28
Recent adopters	1998	Small	6	10

Thus, the first *adopters* of aquaculture in Godilanka were large and big farmers from Rajulu caste, the dominant land owning caste in Godilanka. They had learnt about it from their relatives in the vicinity. Their social position and attitude to take risk had encouraged them to adopt aqua farming. Besides, they also have better educational background. Thus, they became the first adopters of aquaculture in Godilanka. In the first phase, during 1993, about 52 acres of land was converted into aqua (prawn) ponds by 18 landowners. Due to lack of time and non-availability of organised labour had compelled them to engage proclain[2], to convert their agricultural lands into ponds. In their study on Thailand, Flaherty *et al.*, observed that, in the small holder based agrarian structure, the most influential people are often shrimp farmers themselves. Village heads and other members of the local administration are particularly likely to be involved, as they are typically among the largest landowners. (1999:2056)

In the second stage, during 1994-95, seeing the profits earned by the first adopters, who were their close relatives, many large and medium farmers had switched over to aquaculture. During this stage, 26 landowners had converted about 67 acres of land into ponds. They used both proclain and manual labour in preparing the ponds. Manual labour was employed in areas where the proclain could not be used due to space constraints.

In the third stage, between 1996 and 97, quite a few medium and small farmers (18 members) have converted 28 acres of productive agricultural land to prawn farms. Lastly, six small farmers in 1998 had converted about 10 acres of their agricultural land and coconut gardens into ponds. All these farmers had used only manual labour for this conversion.

III

Processes of Aquaculture

The process of aquaculture includes various stages from selection and preparation of the pond to marketing of the produce. The various stages involved in shrimp farming are: 1) site selection and micro level survey[3], 2) Design and construction of farm, 3) Pond preparation, 4) Selective stocking with fast growing seeds, 5) Supplementary feeding and water quality management, and 6) Harvest and marketing. However, the process of aquaculture in Godilanka is not on technically recommended lines, but guided by experience of aqua farmers.

Site Selection

The lands near *Kousiki*, which is the drainage canal, are best suited for shrimp farming because of abundant quantity of saline and irrigation water, and land elevation. Each pond has a tube well for drawing ground water. The aites are of average elevation and are watered and drained with high tides and low tides, respectively. The soil in the bottom of pond is sandy clay-loam to sandy-loam and is most suitable for prawn/shrimp culture. The water used for culture is not checked for ammonia, dissolved iron, hydrogen sulphide and other micro-level elements.

Design and Construction of Pond

The ponds in the Godilanka are both rectangular and square shaped. However, the rectangular ponds are more in number than the square ones. The depth of the ponds is less than one metre and the earth excavated is used as embankment. In the case of marginal, small and medium aqua farmers the same ponds are used as nursery and grow-out pond, while the large farmers maintain separate nursery and grow-out ponds. The ponds consist of structures such as dikes, canals (inlet as well as outlet) and sluice gates, and support facilities such as pumps, aerators, generators, farmhouses, etc. The quantum of facilities depends upon the size of the aqua farm.

Pond Preparation

Pond preparation is carried out prior to stocking of seeds. The bottom of the pond is made flat with one or two long furrows for the shrimp to rest. The bottom of pond is treated with agricultural lime (Calcium carbonate) and dung. The agricultural lime is helpful for the formation of shrimp shell, while the dung aids in the growth of plankton that is beneficial for the maintenance of ecosystem of the pond.

The tank is then filled with water from *Kousiki* with diesel pump-sets. The water is let out for a couple of days as some of it would seep in. Depending on the salinity of the water in the tank, water from the *Kousiki* or the tube well or canal is let into the pond. The farmers or aqua workers in Godilanka physically taste the water for testing its salinity. After a few days, depending on the plankton (plantation, as is locally called) formation, inorganic fertiliser is applied to the tank. The fertiliser helps in growth of 'plantation' on the bottom of the tank.

Selective Stocking

Initially, farmers procure seed on trial basis from the seed businessmen, whose seed tanks are located in Godi, to test their survival ratio. While taking the seed for trial from the seed businessperson, farmers specify their requirements for the crop season. Depending on the survival ratio, aqua farmers stock the seed in their tanks. The seed (post-larval form) thus stocked in a pond is of uniform size. The prawn seed can be either wild or hatchery grown. It is transferred to the ponds in the early hours of the day and evenly distributed over the pond area. Initially, wild prawn seed was procured by farmers from Markalu (fishermen), who collect it on the coast. It is available only during winter season and as such could not meet the demand and also its cost was high. In order to overcome the dearth of seed and its escalating costs, hatcheries were established. Some of the small and marginal aqua farmers still resort to wild prawn seed. However, hatchery seed is preferred by a majority of the aqua farmers as it is cheaper and large quantity of uniform size seed would be available to obtain a uniform size of harvest. Density of stocking in general is inversely proportional to the size of the shrimp at harvest.

Feed and Feeding Practice

Commercial feed for prawn is purchased from Allavaram or Amalapuram market and the feeding pattern observed among Godilanka aqua farmers is of semi-intensive type. Major portion (about 60%) of operation costs goes towards feed. Feeding is an important operation and it is done with lot of care. If proper care is not taken while feeding it will result in not only wastage of feed but it also pollutes the pond. The aqua farmer always checks whether the feed is appropriately administered and consumed. As shrimps go near the dike, feed is broadcasted near the dike. However, care is taken to feed the shrimp in the pond and the feed is broadcasted uniformly for obtaining homogeneous size of the prawns. The farmer reduces the interval between two successive feedings as the shrimp grows in size. Feeding frequency observed is 2 times a day for the first month in which the ration is equally divided between 7 a.m. and 5 p.m.; 3 times for the second month; 4 times for the third month; and 5 times for the fourth and fifth months. The timings of feeding differ as the frequency of feeding increases. In case of feeding after the third month, it is even given in the middle of the night.

After selective stocking, feeding of the seed is done on the basis of some estimation of survival rate. The aqua farmers employ *markalu* (fishermen), who throw the fishing net in to the pond to estimate the survival ratio of prawn and their growth. This facilitates them in feeding the prawn accordingly.

Two months after stocking the seed in the pond, *markalu* are engaged. Till then, farmers feed the prawn based on their understanding and spèculation about the seed. This is purely based on their experience in prawn farming. The farmers on a *manchi roju* (an auspicious day) engage *markalu* for estimating the prawn. The farmers on this day invite their intimate friends and well wishers near the pond. From then on, they regularly employ *markalu* to know about the growth and condition of prawn. The stage of moulting and the growth of the prawn depend on it. During these days farmers advise *markalu* to put net randomly all along the pond and in the centre to have details of the crop.

Every time *markalu* are paid Rs. 30/- to 50/-. This depends on the number of times he throws his net. Aqua farmers always employ *markalu* selectively. They are believed to bring luck to the farmer. *Markalu* working with their relatives are preferred because the farmers feel that they may reveal the details their crop to other farmers for whom he works. There are about 12-15 *markalu* from neighbouring villages who are engaged by the aqua farmers of Godilanka.

Handling of Prawns during and after Harvesting

Harvesting is done when the shrimp reaches marketable size, which is after 4 months of rearing. The preferred days of harvesting are four days before or eight days after full and new moon period, respectively. Immediately after these days the shell of the prawn is lost due to moulting and it weighs less, and it is also difficult to catch the harvest. Normally the size of prawn at harvest varies between is 20-30 gms.

Markalu and casual labour are engaged on these days. The time lapse between the harvest and post-harvest handling of prawn is important. The prawn harvested is given a quick wash to remove dirt. After this, they are kept in iceboxes, supplied by the marketing agents, as immediate chilling reduces bacterial growth and prevents decomposition and change of colour of prawns. In some cases they are beheaded and in others they are kept intact. Both the 'head-on' as well as beheaded shrimps are stacked in crates with alternate layers of crushed ice and transported to the nearest processing plant. If the harvested prawn is kept longer outside the iceboxes its colour changes and it gets a low price to the farmer. Even this price depends on the marketing agent, who may or may not buy them. They are normally processed within 24 hours of harvesting.

IV

Relations of Production

The relations of production in aqua farming essentially deal with the relations between the owners and workers. As in the case of agriculture, in aquaculture, the non-owners are tenants and *aqua workers* are hired labourers (Chart 4.1). Continuation of the tenant depends on the nature of relationship between the tenant and the owner, the margins of profit and availability of credit.

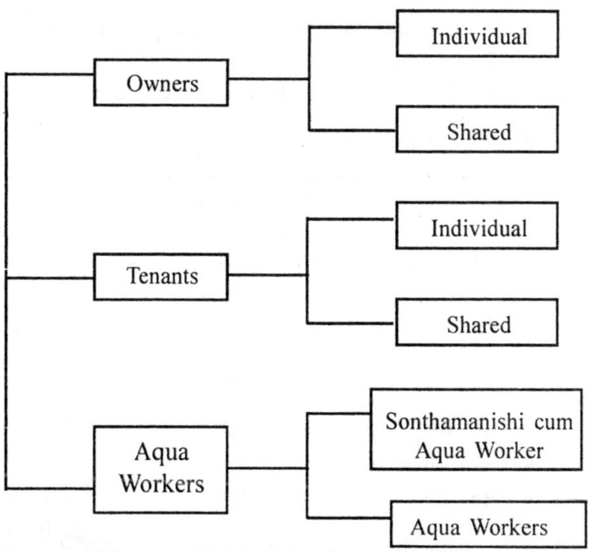

CHART 4.1: SOCIAL ORGANISATION OF AQUACULTURE

There are two categories of aqua tenants, individual and joint (sharing), as in agriculture. The tenants who can afford to pay in advance the agreed amount to the landowner cultivate individually, while those who can not afford to pay individually resort to joint farming with some close relatives, who share both the costs and profits. There are no conditions of leasing, except payment of agreed amount of cash at the time of taking the pond for lease for a specific period of cultivation. This type of leasing is similar to *gutta vyavasayam* in agriculture, except for the fact that the payments are made at the end in agriculture. In case of aqua workers, the conditions of work and wages are negotiated in the beginning and his continuation depends on the nature of relationship between him and the aqua farmer.

Owners

Based on the ownership of area and extent of aquaculture in the village, aqua farmers can be classified more or less in a similar way as in agriculture. They are 1) marginal aqua farmers, 2) small aqua farmers, 3) medium aqua farmers, and 4) large aqua farmers. Marginal aqua farmers are those who cultivate prawn in less than 1 acre pond. The aqua farmers between 1.1 acres and 2.5 acres ponds are treated as small aqua farmers. The medium aqua farmers maintain aqua farms between 2.6 acres to 5 acres. Large aqua farmers own the largest area under aqua farming in Godilanka and the extent of their culture is above 5 acres.

Table 4.4: Type and category of both aqua farmers and farmers

Category of Aquafarmer	Aqua farmers	Aqua farmers-cum-agriculturists	Total
Marginal	5 (26.3)	14 (73.7)	19 (100) (29.7)
Small	3 (12.0)	22 (88.0)	25 (100) (39.1)
Medium	4 (30.8)	9 (69.2)	13 (100) (20.3)
Large	4 (57.1)	3 (42.9)	7 (100) (10.9)
Total	16 (25.0)	48(75.0)	64 (100) (100)

Note: Figures in parentheses are percentages.

There are two types of aqua farmers, exclusive aqua farmers and aqua farmers-cum-agriculturists. The small and marginal aqua farmers constitute two third of aqua farmers and among them a majority belong to aqua farmers-cum-agriculturists. Between both these categories marginal aqua farmers have a higher proportion of exclusive aqua farmers than the small aqua farmers do.

In contrast, large aqua farmers have higher proportion of aqua farmers than aqua farmers-cum-agriculturists do. It is important to note that there is a gradual increase in the proportion of exclusive aqua farmers from small farmer category to large aqua farmer category. Thus the higher the category of farmers the lesser the chance of being involved in agriculture, besides aquaculture (Table 4.4).

The owners of aqua farms in Godilanka exclusively belong to Rajulu, as they are also owners of most of the agricultural land. Most of the aquaculture in this area is controlled and managed by either their kin/caste members in the region. Risk taking attitude, knowledge of aqua farming, easy access to inputs and availability of credit are important factors responsible for their venturing into aqua farming. Many farmers of this village also have lands in the neighbouring villages within this Mandal. Most of these lands, depending on the accessibility and availability of backwaters, were converted into aqua ponds.

In Godilanka, as noted earlier, exclusively Rajulu do aquaculture. It is essential to note that the aqua farmers are highly enterprising farmers and in order to understand a profile of the aqua farmers there is a need to understand their educational and age backgrounds.

The aqua farmers in Godilanka predominantly belong to the 25-35 years age group (62.9 %), followed by the 36-45 years. While there are none in the age group of above fifty, there are very few in the age groups of 46-50 (3.2%) and below 25 (8.1%). This shows that the younger and middle age groups are more inclined to aquaculture as compared to the others (Table 4.5).

Most of the aqua farmers are educated up to inter or had technical education (33%), like ITI or polytechnic, followed by high school and college and above levels of education. It is pertinent to note that, unlike the agriculture farmers, illiteracy among aqua farmers is minimal and most of them are better educated (Table 4.6).

Table 4.5: Aquafarmer-Age

Age group	Number
Below 25	5(8.1)
26-35	39 (62.9)
36-45	16 (25.8)
46-50	2 (3.2)
50 and above	-
Total	62 (100)

Note: Figures in parentheses are percentages.

Table 4.6: Aqua Farmers and Education

Level of Education	No. of farmers
Illiterate	3 (4.8)
High school	18 (29.0)
Inter/ITI/Polytechnic	21 (33.9)
College and above	20 (32.3)
Total	62 (100)

Note: Figures in parentheses are percentages.

In Godilanka aqua farming is practised either by individuals or jointly with others (sharing). The total area under aquaculture is 154.72 acres. Of this individuals cultivate about 101.72 (65.7%) acres of aqua ponds and the rest is done jointly. The sharing type of aqua farming is noticed mostly among close relatives.

Aqua Farming by Individuals

The individual aqua farmers include owners and tenants. The individuals involved in this type of farming have the capability to meet the requirements for the enterprise. It involves investment to meet capital costs (seed, feed, etc.) and labour for the crop season. An individual has to raise capital for investment or borrow from other sources. Aqua farming involves unforeseen consequences that may arise due to unfavourable climatic conditions, spread of disease to the prawn, etc. The affects of this are more adverse to the individual tenants as compared to individual owners.

Aqua Farming by Sharing

In Godilanka, farming by sharing is noticed among Rajulu. Individuals involved in this type of farming are close relatives (consanguineal or affinal). There are two types of sharing among the aqua farmers, owner sharecropping and tenant share cropping. In case of owners they might do share cropping with a relative (s) or might lease out to tenants who, in turn, will cultivate it by sharecropping with a relative (s). It is important to note that both types of aqua farmers are from Rajulu caste only. In Godilanka 39 acres of tanks are under owner sharecropping and 14 acres of tanks are under tenant sharecropping.

In the owner-shared aqua farming two or three relatives are usually involved. This kind of sharecropping depends on the resources of the persons involved and the number of persons involved depends on the extent of aqua pond. The individuals involved in this type of aqua farming are those who possess land but have no capital. Here they need support from close relatives, who would either invest money or oversee the cultivation. Close relatives are regarded as more trustworthy in a highly capital-intensive enterprise. The closer the relationship the greater will be the trustworthiness. This is clearly demonstrated in the following types of sharing with different types of relatives.

The degree of relation also is an aspect to be understood in share-cropping in aquaculture. Quite a significant amount of aqua land is cultivated jointly with consanguineal relatives (67.1%) as compared with affinal relatives (32.9%). Among the consanguineal relatives, persons from the same family (41%), brothers, are more involved in share cropping than first or second cousins (37%) or distant relatives from the same lineage (22%). Among the affinal relatives sharecropping is undertaken mostly with *menalludu* (sister's son) (44.9%), i.e., between *menamama* (mothers brother) and *menalludu* (sisters son) and with *bavamaridi* (wife's brother) (25.8%), i.e., between *bava* (sister's husband) and *bavamaridi*. Among the others, sharecropping with *alludu* (son-in-law) (17.2%), i.e., between *mama* (father-in-law) and *alludu*, is higher than with other distant affinal relatives (11.9%), such as *sadugudu* (son-in-law's father or daughter-in-law's father), *todualludu* (co-sons-in-law), etc. In share cropping of aquaculture the land used for pond, capital involved and the labour are not similar. Thus, the returns always depend on their initial investments in terms of land, labour, and capital.

The following case of three brothers, who are doing aqua farming in 1.7 acres of aqua pond, reveals the nature of sharecropping in aquaculture in Godilanka. Prior to the introduction of aquaculture in Godilanka three brothers R. Gopal Raju, Bheema Raju, and Kesava Raju were doing *paalu vyavasayam* in 4 acres jointly. They owned 2 acres of coconut garden jointly, which was inherited from their father. In 1995 the land in which they were doing *paalu* was converted to aqua tanks and they were displaced from their work. The succeeding year they had resorted to *gutta vyavasayam*. In the next year they planned to undertake aquaculture and managed the required credit. Thus, they left agriculture and leased-in 1.7 acres of tank and jointly did aquaculture. The landowner, with whom they did *paalu* for years, had arranged for feed from a distributor from Allavaram, a native of Godilanka, who is also known to them. The brothers had contributed and worked equally on the aqua farm. They looked after all the activities without employing any aqua worker. In the first crop the profit was adjusted against lease, diesel pump-set and other inputs, which they had to invest in. In the second crop each of them earned a profit of Rs. 30,000/-. They continued practising aqua farming in the same manner.

It is important to note that consanguineal relationship (being brothers) is the basic binding factor for them to engage in aquaculture. This was also the same during *paalu vyavasayam* in agriculture. In addition to this, living together in their ancestral house with their property undivided has also been the contributing factor for them to work together. Further the area under agriculture was reduced due to aquaculture and, hence, the brothers had limited choice between agriculture and aquaculture. Their opting for leased-in aquaculture turned out to be a successful effort.

In case of siblings who possess the undivided landed property continue to have equal access to it. Any capital investment on the lands will be shared equally. The labour employed on the tank will be from either their existing labour in agriculture (*sonthamanishi*) or contracted separately for this purpose. They share the costs of labour equally. While sharing profits, costs incurred on the inputs and labour will be deducted and the remaining will be shared equally.

If the sharing is with cousins (first or second) or other relatives, the pond may belong to any one of them. The investments and the returns are shared according to an understanding between or among them. Aqua worker is either contracted or one among them may work as aqua worker to look after day-to-day activities.

Thus in share cropping involving consanguineal relatives, siblings have greater affinity than the cousins do. Among the cousins, the second cousins and distant relatives have lesser affinity than the first cousins. Though the relationship among distant relatives operates mostly on economic lines, kinship continues to operate in the continuation and maintenance of collective farming of prawn by them.

This case study of share cropping between *menamama* and *menalludu* indicates the nature of sharecropping in aquaculture involving affinal relatives. Alluri Sekhar Raju aged 45 years, studied up to intermediate. He is a big farmer with 10 acres of paddy and 12 acres of coconut garden. In 1994 he had converted 4 acres of his agricultural land into aqua ponds. During the first year he reaped a good harvest and the second year crop suffered due to high density stocking. This led to dissolved oxygen (DO) problem affecting the prawn yield, which hardly compensated his investments. Following this experience, he had stocked his pond with average density (20,000 seed per acre) in the next year, which resulted in good yield.

Sekhar Raju's sister was married to R. Venkateswara Raju, belonging to the same age group. He is a medium farmer with 2 acres of paddy and 4 acres of coconut plantation. His son, Srinivas Varma, 22 year old, is a commerce graduate and had little interest in paddy cultivation. He preferred aquaculture due to its higher profit margins. Besides, many of his peers were also involved in it. Venkateswara Raju's paddy land had no access to saline water.

Since Varma approached his *menamama* Sekhar Raju for taking up aquaculture. Conceeding the request of his nephew, Sekhar Raju had converted another acre of agricultural land into aqua pond with a condition that Varma has to take care of the pond by himself. This was because Sekhar Raju's aqua farms were far away from this aqua pond and it would not be possible for him to involve in aquaculture in the new aqua pond with his nephew. The investments between *menamama* and *menalludu* were in the ratio of 1:3 and Sekhar Raju arranged even this through a moneylender. The feed, seed and other technicalities, except Aqua worker, were acquired under the guidance of Sekhar Raju. They got a good harvest in the first crop and after deduction of costs of inputs and labour, Varma got a profit of Rs. 20,000/-. They continued their collaboration for the subsequent seasons also. In the second season, in addition to aquaculture, Varma also took up seed business.

It is because of his relationship, Sekhar Raju had invested in digging the pond and arranged credit to his *menalludu.* He had also taken care for seed, feed, etc. Varma had to contract an aqua worker and pay his wages, monitor the pond and manage the requirements on the pond. Thus in share aqua farming relationships play a vital role in helping their relatives to improve their economic status.

In sharecropping with affinal relatives, unlike the consanguineal relatives, land, capital and labour are unequal. Land belongs to one, while the others in this type of a venture take care of the labour and capital. In this, one of the individuals involved is of lower economic status and the other helps him to come up economically. Thus affinal or potential affinal relationship plays an important role in the upliftment of persons or families with lower economic standing. As observed earlier, in Godilanka the bond between *menamama* and *menalludu* is the strongest when compared to *bava-bavamaridi* and *mama-alludu.* The *menamama*'s relationship with *menalludu* is stronger because of two reasons, one is that he is a potential son-in-law and secondly, *menalludu* has customary rights and obligations to his *menamama* and *vice versa.*

Apart from relatives, close friends also involve in share aqua farming. The friends may or may not be from the same caste. The case of *Nadimpalli* Linga Raju and his friends from his own caste, who undertook aquaculture jointly, indicates the role of caste as well as friendship in share aqua farming.

Nadimpalli Linga Raju, aged 32, is a graduate now residing in Amalapuram. He has his lands and aqua ponds in Godilanka and he has been busy with them. *Budaraju* Satya Varma, a 36 years old graduate working as a clerk in Amalapuram was a friend of Linga Raju. Ramakrishna Raju, another friend of Linga Raju, aged 30 years, has done ITI at Bhimavaram and later worked at Hyderabad. Godilanka was his maternal grandmother's place and his Grandfather had educated him. During every vacation, he used to come to Godilanka and stay with his grandparents.

All the three friends are from Rajulu caste and wanted to do aquaculture jointly. Both Linga Raju and Satya Varma were busy in their own activities and they had little time to spend on this venture. They came to an understanding with Ramakrishna Raju that they will invest and he had to only supervise the aqua farm for which he would get 10% of the profits. They had leased-in 2.3 acres of aqua pond and both of them invested in the ratio of 50:50. After paying ten per cent of the profits to Ramakrishna Raju for rendering his services, they shared the rest of the profit between them equally. Ramakrishna Raju got Rs. 30,000/- as part of his ten per cent share.

The following case of leasing-in of aqua pond by friends from different castes as well as relatives together throws light on the role of friendship on share aqua farming and its difference can be seen with the above case. Rudraraju Ashok Raju, prior to the introduction of aquaculture in Godilanka was involved in buying and selling of paddy to mills in Amalapuram. In 1994 he worked on one of his relative's aqua farm for which he received Rs. 20,000/- as part of his share for supervision. In 1995 he undertook jointly aqua farming with three of his relatives. All of them invested equally and converted their 2.5 acres of agricultural land into aqua pond. His relatives looked after feed, seed, etc., as they knew feed agents. First crop was adjusted towards inputs. In the next crop each one had got a profit of Rs. 25, 000/-. They continued the same for the second year. They reaped profits for the first crop during this year. However, they lost the second crop at '50 count' stage and the harvest could only meet major part of their expenses. After this catastrophic experience all his partners in the venture had withdrawn.

Now after two years of aqua farming he had accumulated some money and leased-in aqua farm individually (2 acres) in 1997. He has a friend, P. Srinivasa Rao, from Komati caste, who leased-in a rice-mill in Godilanka, who was also interested in aqua farming. Thus they got together first to do joint aqua farming. Ashok Raju had invited his *bavamaridi* residing in a village near Ramachandrapuram, about 50 Kms from Godilanka, to join them in aqua farming. His *bavamaridi* had invested one fourth, while both himself and his Komati friend invested the remaining amount equally. Thus they have farmed 2 acres by sharefarming. Both Ashok Raju and his Komati friend monitored the crop. The first crop was adjusted towards the inputs and the second crop had given them huge profits, which was divided according to the ratio of their investments. After two crops, Sambasiva Raju an[d] Komati friend got a profit of a lakh each and his b[...] around 65,000/-.

Among these landowners, aqua farming by sharing, who possess land and have insufficient capital or do not want to take risk, but venture with active support of relatives. The relatives join here as partners and share the cost of inputs. The ratio of sharing of inputs and profits are worked out according to a prior agreement. In many cases of this type the owner works on the farm.

This type of share cropping has some similarity in its essence with *paalu vyavasayam* in agriculture, though there are some basic differences. In agriculture, for procurement of labour the landowner leases-out land for *paalu* to tenants and he never works on the land. In contrast, the people participating in aqua share farming are his relatives. The inputs and profits are jointly shared and the owner most of the times works on the pond. His relatives or friends, who are partners, may work and invest or may only invest along with him in the venture. Thus in a way an essence of *paalu* is noticed among the landowners sharing in aqua farming.

Mobility of an Aqua Farmer

The emergence of aquaculture in Godilanka has been attracting many landowners, who earlier had migrated out of the village while their lands were supervised by some of their relatives, either brothers or others. The younger generation of landowners after their education preferred to work as white-collar workers in government or industry in towns or cities. These persons returned to Godilanka after the emergence of aquaculture to take up aqua farming as they learnt from their relatives that the returns were quite high in aqua farming.

These members first involve themselves in seed business as they lack experience. They are not encouraged by their parents or other elder members to directly take up aqua farming by converting agricultural lands. Starting of seed business needs some investment but involves little risk. Their family members or relatives encourage them by buying seed from them. Meanwhile, they also find a few new farmers to sell their seed. In this process they acquire the knowledge of seed business from their elders or peers who were involved in this business earlier. They will continue for a year or two in seed business and with the profits in seed business and with some additional investment they think of participating in aquaculture jointly by share farming with relatives or peer group members, depending on their confidence in them. They continue this type of share farming for one to three years, till they earn sufficient money and confidence to start individual aqua farming. During the years of share farming they obtain necessary information and develop net works necessary for aqua farming. After this, they engage in individual aqua farming either on their own ponds or on leased-in ponds. Flaherty *et al.*, observed that, "...lucrative shrimp farming can be compared to rice cultivation, and ...large numbers of farmers who are able to raise the investment capital are now willing to take a gamble on raising shrimp. In provinces where rice farmers are unwilling or unable to invest themselves, there is ample opportunity for leasing paddy land to outside investors at rents that greatly exceed what they could obtain growing rice, or for a share of the profits" (1999: 2050).

In Godilanka the land under aquaculture is limited and there is an increase in the number of aqua farmers, especially among the younger members. This has led to a greater demand for leasing of aqua farms. As a result, the leasing rates of aqua farms are being enhanced every year. Thus, there is a greater swelling of ranks to the category of aqua tenants.

V

The persons who lease-out aquaculture ponds in Godilanka, as in the case of agriculture, are widows, employees and large farmers. These individuals either do not have manpower to monitor ponds or fear that since aquaculture requires huge investment, they may not be able to withstand the losses if the crops fail. The lands belonging to the above persons located in and around *Kousiki* are leased out to Rajulu caste tenants.

Tenants

They are a category of individuals who always lease-in aqua farms from their owners. Leasing is based purely on economic considerations. The tenants lease-in land and convert it into aqua ponds or directly take aqua ponds on lease for a specific period of time, usually one to three years. If the land is taken for lease and converted to aqua pond, the costs of digging the pond are to be borne by the tenant. His investment for digging the pond is adjusted towards the lease amount in the first year and he can raise crop during this period without any payment. The leasing-in of the aqua ponds depends on the aspects like investment and returns in previous crop season, availability of credit, market price of the prawn and individual capability to invest huge capital. Above all, it depends on the knowledge and understanding of the prawn farming of the tenants.

The tenants of the prawn farms, as in case of agriculture, can be grouped into tenants, and owner-cum-tenant. Among the tenants 17 individuals are pure tenants and 4 persons are owners-cum-tenants. About 34 acres of land is under aqua tenant farming.

The aqua tenants in Godilanka can be categorised into two types, individual and joint aqua leasing. In the later case, two or more persons (less than five) come together and undertake aqua farming in leased-in lands. As in case of owner share farmers, tenants are also peer group members from the same caste within the village. The persons involved in joint aqua leasing invest either equally or differentially depending on their economic condition. The returns are shared according to their investment. In some cases, an individual who is economically weak and can not invest will look after aquaculture from preparation of pond to harvest of crop, including working as aqua worker. These individuals receive some amount mutually agreed upon for their services of supervising and managing the crop, as in case of owner share farming observed earlier. There are only a few cases of this type in Godilanka.

Lease Rate and Conditions

After the introduction of aquaculture in 1993 in Godilanka, the ponds were first leased-out in 1994. The leasing rates per acre per year were initially Rs. 10,000/-. The pond was dug by the owner and was leased-out to the tenant. In general, many tenants leased-in ponds for three years and the lease amount was given to the owner after the harvest in the initial years. In 1996 many tenants have converted agricultural lands as well as coconut gardens into aqua ponds. The money invested for this conversion was adjusted towards the first year's lease amount.

The lease amount depends on, the last crop yield, access to ground water and diesel pump-set at the aqua farm, drainage facility and general yield of crop in the village. The lease amount has almost doubled by 1996 and ranged between Rs. 15,000/- to Rs. 18,000/-. Half of the money was paid at the time of execution of the documents and the rest paid after the harvest of prawn. The second crop of 1997 was not very encouraging. It was partly due to the spread of diseases to the prawn and low market price. A few tenants had postponed their payment of lease amount. This has led to collecting the total lease amount by the landowners at the time of signing the documents. In 1998 the lease amount ranged between Rs. 25,000/- and Rs. 30,000/- per acre per year. This increase in leasing rates are because of the competition among Rajulu to procure ponds for aqua farming, which are promising in terms of profit as compared to agriculture (Table 4.7).

Table 4.7: Lease Rates of Aqua farms

Year	Lease Amount (in Rupees)	Mode of Payment
1993	-	-
1994	10,000	After the crop harvest
1995	10,000- 15,000	After the crop harvest
1996	15,000- 18,000	Before and after the harvest (50:50)
1997	18,000- 20,000	Before and after the harvest (50:50)
1998	20,000- 28,000	Before the commencement of the crop

The lease rates in Godilanka are higher than the corresponding rates in the surrounding villages. This is because in Godilanka crop yields were good while the crops in the neighbouring villages were low. Caste and primordial relationships always have a bearing on the leasing contract as well as mode of payment. Depending on the degree of relationship the lease rate and the mode of payment varies. The relatives consider some relaxation in the conditions of payment of lease money.

In 1996 Rudraraju Appala Raju, aged 42 years, had switched over to aquaculture by converting 2.6 acres of his land. He reaped a bounty of harvests for the first two crops followed by a very low crop yield due to disease to prawn and again another good crop in the next season. This has lured him to expand the extent of aqua farming. In 1998 he leased-in 1.8 acres of land from R. Narayana Raju.

Narayana Raju, a 60 years old man, has a son who has been involved in aquaculture (5 acres leased-in) since two years in an adjacent village, Gundupudi. Narayana Raju has a small patch of coconut garden near *Kousiki*, which was surrounded by aqua tanks on two sides. As a result, the yield has decreased substantially and even some trees died due to seepage of saline water from the adjacent tanks. He had no option but convert it to aquaculture or lease-out for the same purpose. He had no one to take care of it and due to his old age and failing health, he had decided to lease-out this land to Appala Raju. The lease period was for three years. The condition of contract was that the cost for digging was to be borne by the tenant, which would be adjusted against the lease amount for the first year. After the end of lease period, the continuation of leasing depends on their understanding and the prevailing lease rates.

After leasing-in the land, Appala Raju had converted it to aqua pond by employing manual labour. As the profits in aqua farming were higher Appala Raju's father also joined him in aqua farming. He started attending all the activities of pond conversion. As per agreement the first year's lease was waived. This encouraged them to lease-out their 2 acres of agricultural land to *gutta vyavasayam* to totally concentrate on aqua farming.

There are many cases similar to the above. The above case indicates a process that has been taking place in the changing agrarian scenario due to the expansion of aquaculture. Many farmers, who would otherwise be in agriculture, have converted or converting agricultural lands as well as coconut gardens for monetary considerations. Some others lease-in land or coconut gardens and convert them to aqua farms. The earlier land and labour relations are adapting to the changed circumstances.

VI

The category of labour employed in aquaculture is aqua workers. In addition to this category, a few casual labourers are employed in different activities as and when required on the aqua farms.

Aqua Worker

The aqua worker is a hired labourer engaged in aqua farming. This is similar to *sonthamanishi* in agriculture. Locally they are known as *cheruvu meeda manishi* (person staying on pond). He is employed just before the crop season starts and is a key labourer who will be involved in the preparation of pond; and in feeding, monitoring and harvesting of the crop. In the preparation of the pond, the aqua worker treats the bottom of pond with agricultural lime (Calcium carbonate) and dung. He then fills the tank with water from *kousiki* to the required level by running diesel pump-set. After a few days fertiliser is applied to the tank. This is done to grow moss on the bottom of the tank, which prevents the water getting heated up due to sun light, since it acts as a bad conductor of heat from the ground to the water. Application of dung helps in the growth of plankton that is useful for the growth of prawn and maintenance of the ecosystem of the pond. Aqua worker feeds the seed (post-larval form). Every fortnight he treats the pond either with dolomite or agricultural lime, which helps in the moulting of prawn. In the early stages he feeds the prawn fewer number of times and as the prawns move to the advanced stages he has to feed them more frequently, including during late nights. He maintains the quality of water as directed by the technician and changes water depending on the growth of the prawn. He also arranges labour on the harvest days.

Daily Routine of an Aqua Worker

Early in the morning he goes to the pond and feeds the prawn. Later he returns home around 9 a.m. and takes food. He again returns to the pond to oversee that birds, water ducks, etc., do not prey on the prawn in the pond. His master visits and examines the pond, and instructs the worker to repair the bunds or check the water-lets that draw water in or out of the pond. At about 11 o'clock the worker feeds the prawn and keeps a watch on the pond. He goes for lunch in the afternoon and returns by 2 p.m. At 4 p.m. his master visits and inspects the pond and water level in the pond. Following the instructions of the master, the aqua worker operates the pump-set to fill the water to the required level in the pond. For every two hours he goes round the pond and meticulously observes the health of the prawns and the pond condition, and reports this to his master as and when he comes. At 7 p.m. he again feeds the prawn and leaves for his home.

In the advanced stages of prawn count (50 count and below), the aqua worker has to stay on the pond day in and day out. His master comes and relieves him for food and the aqua worker has to return from home quickly. He even sleeps near the pond during nights as the crop nears harvesting stage. Some of the aqua workers express their discontentment because aquaculture leaves no time for their family life.

Wages

The aqua workers of Godilanka are paid a monthly wage. The wages vary from Rs. 900/- to Rs. 1200/-. This is more or less similar to the wages paid to a *sonthamanishi* in agriculture. Before being contracted for the crop season, the aqua workers are advanced an interest free consumption loan so as to keep him tied for the whole crop season. If the aqua worker wants to leave in between, which he can not do, he has to provide a substitute. The aqua farmer gives aqua workers an interest-free credit, which ranges from Rs. 3,000/- to Rs. 5,000/- per individual.

Table 4.8: Consumption loan received by category (in Rs.)

Type of Labour	3000	4000	5000	More than 5000	Total
Aqua worker	14 (33.3)	13 (30.2)	12 (27.9)	4 (9.3)	43 (100) (67.2)
Sonthamanishi and/aqua worker	3 (14.3)	9 (42.6)	7 (33.3)	2 (9.5)	21 (100) (32.8)
Total	17 (26.6)	22 (34.4)	19 (29.7)	6 (9.4)	64 (100)

Figures in parentheses are percentages.

The number of aqua workers employed by a pond owner depends on the extent of pond. Usually, for every 2.5 acres of pond area one aqua worker is employed/contracted. In case of earlier *sonthamanishis*, they continue to work in the same capacity. If the landowner has both agriculture and aquaculture, In this case, the *sonthamanishi* will work as aqua worker on the aqua farm and switches over to his *sonthamanishi* role during agricultural season.

Table 4.8.1: Consumption loan received by Aqua worker (in Rs.)

Caste	3000	4000	5000	More than 5000	Total
Rajulu	-	-	1(25)	3(75)	4(100)
Settibalija	-	2(33)	3(50)	1(17)	6(100)
Mala	8(36)	8(36)	6(27)	-	22(100)
Madiga	6(55)	3(27)	2(18)	-	11(100)
Total	14 (33.3)	13 (30.2)	12 (27.9)	4 (9.3)	43 (100)

Figures in parentheses are percentages.

If the landowner has aqua farm only, depending on the extent of pond, he retains his earlier *sonthamanishi* and in this case he works as an aqua worker. Roughly, one third of aqua workers are working both as '*sonthamanishi* and/aqua workers'.

Table 4.8.2: Consumption loan received by Sonthamanishi-cum-Aqua worker (in Rs.)

Caste	3000	4000	5000	More than 5000	Total
Rajulu	-	-	1(100)	-	1(100)
Settibalija	-	2(50)	1(25)	1(25)	4(100)
Mala	1(10)	4(40)	4(40)	1(10)	10(100)
Madiga	2(33)	3(50)	1(17)	-	6(100)
Total	3 (14.3)	9 (42.6)	7 (33.3)	2 (9.5)	21 (100)

Figures in parentheses are percentages.

Majority of the aqua workers is given an average amount of 3,000 to 5,000 consumption loan at the time of contract, depending on the length of service under the aqua owner. However, a *sonthamanishi* and/ aqua worker gets higher credit than an aqua worker does because of the length of service under the land and aqua owner. The loan amount increases with the no of years one has worked with a landowner, either in agriculture or aquaculture (Table 4.10). Further, besides the length of service under a owner, the caste of the aqua worker or '*sonthamanishi* and/aqua worker' also matters in receiving the quantum of credit.

'Private' labour is also engaged in work as the situation demands and they had to be paid after a day's work. Some times, if a 'private' labourer works for a longer period of time on a single pond, he is paid the wages all at once in the end. Unlike the aqua worker, he can quit the job whenever he wants.

After the emergence of aquaculture in Godilanka the aqua workers have emerged as a category of work force. They were drawn from among the earlier workers in agriculture within the village. The emergence of aquaculture did not lead to any out-migration of labour or unemployment among them. It is pertinent to note that the number of days of work has come down to a certain extent, but it balanced out in terms of higher wage rates. Both the aqua and agricultural workers now feel that they have better opportunities to improve their living standards than earlier.

However, it is significant note that aquaculture has affected the workforce among women. Seasonally women now go to the neighbouring villages during harvest and transplantation days during the agricultural season, besides working as labourers at the time of prawn harvesting.

It is observed that most of the aqua workers in Godilanka were earlier working as *paleru* and *sonthamanishi*. Nearly three fourths of the aqua workers are from Mala and Madiga castes. Quite a few aqua workers among Mala and Madiga were earlier working as *paleru*, while in case of Settibalija none were drawn from this category. Thus we find a majority among Rajulu and Settibalija came from *sonthamanishi* background (Table 4.9).

The following case illustrates the adaptation of *sonthamanishis* to aquaculture. Prior to aquaculture, *Penumala* Sudhakar, 35 years of age, was working as *sonthamanishi* with R. KanakaRaju. His father also worked as *sonthamanishi* with KanakaRaju for more than 7 years and was given *paalu vyavasayam*. After marriage he joined as *sonthamanishi* with the same landowner. Some of the agricultural lands were converted to aqua tanks by the landowner KanakaRaju and Sudhakar was shifted to aqua work. Now Sudhakar oversees and arranges the required labour during peak aquaculture season. During this period 'private' labour are employed on the pond by the pond owner. From the beginning of aquaculture he has been involved in aqua farming under KanakaRaju. He is paid 5 *manelu* of paddy and Rs. 5/- *beta* as wages per day. In addition to this, he receives customary payments usually given to a *sonthamanishi*.

Table 4.9: Earlier Occupations of Aqua Workers by Caste

Caste	Earlier Occupation			Total
	Paleru	*Sontha-manishi.*	'Private'	
Rajulu	-	14 (93.3)	1 (6.7)	15 (100)
Settibalija	-	4 (100)	-	4 (100)
Mala	8 (27.6)	18 (62.1)	3 (10.3)	29 (100)
Madiga	6 (37.5)	7 (43.7)	3 (18.8)	16 (100)
Total	14 (21.9)	43(67.2)	7 (10.9)	64 (100)

Figures in parentheses are percentages.

The case of an educated Mala caste member who was working as a daily wage labourer in an electrical firm in Hyderabad illustrates the pull factor in migration due to the emergence of aquaculture.

Settibathula Shoban is a 24 years old Mala caste member from Linepet, who had studied upto 10th class. Even after several attempts he could not pass SSC examination. However, he was not inclined to work as agricultural labourer and preferred to work in any industry as a worker. Being educated, he felt that working as an agricultural labourer was below the dignity. So, he went to Hyderabad in 1994 in search of a job, where some of his caste members from Godilanka helped him to secure a job as a daily wage labourer in an electrical firm. He was paid Rs. 40/- daily. During his stay in Hyderabad he was living with his friends and more than half of his income was spent on food and rent. He found it very difficult to sustain himself with meagre wages after two years. After the emergence of aquaculture wage structure has increased in Godilanka. This has attracted him to return to Godilanka in 1996 and started working as aqua worker with a monthly wage of Rs. 1000/- initially in the first year, which was enhanced to Rs. 1200/- in the next year.

Living in the midst of his relatives along with his family members, increased wages and a risk free life encouraged Shoban to return to Godilanka. Like Shoban there are about a dozen people who had gone to cities and towns in search of employment and after a couple of years have returned to the village on similar grounds.

VII

Aquaculture involves large amounts of investment, which is required in every stage of work, right from conversion of land to pond, purchase of diesel engine pump-sets, buying of seed and feed, etc. Aqua farmers, irrespective of category, need money for aqua farming and for this they depend on formal institutions or informal credit agencies or individuals. The Godilanka aqua farmers mostly received institutional credit from PACS and informal credit from moneylenders. The President and the committee members of PACS are from Rajulu caste. More importantly, due to its location in Godilanka and its organisational control by the local Rajulu caste persons, processing of credit has become easier and quicker for the local aqua farmers.

Table 4.10: PACS loans for Aquaculture

Year	Loan Amount (in '000)					Total
	< 25	25- 50	50- 100	1- 150	> 150	
1993-94 (FT)	-	10	18 + 1*	10	1	40
1994-95 (FT)	2	11	6	1	-	22
1995-96 (FT)	1	2	2	1	-	6
1996-97(FTR)	6	8	14	3	-	31
1997-98(FTR)	7	8	4	-	-	19
Total	9 (9)	32 (33)	40 (41)	16 (16)	1 (1)	98 (100)

* Loan taken by Settibalija caste.
Note: FT = fish tank; FTR = fish tank repairs.

In principle, there is no provision for issuing loans for prawn aquaculture in PACS, but there is a provision for development of fish tank and its upkeep. This is a long-term loan, which should be cleared in 8 years. The loans were advanced against mortgage of collateral property, which are usually agricultural lands. An inspector who values the worth of the land comes from DCCB, Kakinada, and inspects the land. Following his recommendations, which depends on the value of the land mortgaged, the loan amount applied for would be sanctioned based on the eligibility. All the farmers in Godilanka had applied for fish tank loans and managed to get them sanctioned, which they used it for prawn farming. The loan amount granted was 60% to 75% of the value of the property.

FT- Fish tank, FTR- Fish tank repairs

There are four aqua farmers who had borrowed money from a commercial bank from Amalapuram in 1994. The loan amount ranged between Rs. 50,000 to 75,000/- with an annual interest rate of 14%. In the same year another farmer got MPDEA loan for Rs. 1.44 lakhs for aquaculture. This loan was released through PACS with a subsidy of Rs. 34,500/-. Subsidy was given by MPDEA basically to encourage aquaculture in the region.

In addition to formal credit, many small and marginal aqua farmers borrowed money form informal sources, mainly from moneylenders, for investment in aquaculture. The moneylender advances credit only to those aqua farmers who have property either in the form of house or agricultural land.

Table 4.11: Credit for aqua farming (1997-98)

Farmer type	PACS	Moneylender	Bank	Relatives	Total
Marginal	3	22	-	11	36
Small	8	15	-	2	25
Medium	8	3	-	-	11
Large	-	-	-	-	-
Total	19	40	-	13	72

As observed by some scholars (Jodhka, 1995), local power structure interacts with and influences the functioning of PACS. The organisation and functioning of PACS has further strengthened the already dominant and powerful individuals/sections in Godilanka. As a result, the landless, small and marginal have little access to the institutional sources of credit. Thus, they continue to depend on informal sources like moneylenders, *Sahukars*, etc., (Table 4.11).

VIII

Marketing Chains

The prawn farmed in the aqua ponds has to be marketed. This is essential because the prawn reared is costly and chiefly produced to meet the demand of international market. To facilitate its marketing there is an established market chain. The aqua farmers before starting the pond with the help of their relatives, who are already in aqua farming, approach the feed distributors, who mostly belong to Rajulu Caste. The feed distributors are located in Allavaram and Amalapuram. There are many seed distributors and each one of them represents a company. Depending on the affinity towards a particular brand, the aqua farmers order their feed requirements with the distributors of those brands who supply the feed against some security and on a pre-condition that the harvest has to be marketed through him. There are many marketing companies in Amalapuram and Ravulapalem run, by and large, by Rajulu caste persons, who buy the harvested prawn and sell it to processing units or to their company outlets. There are also certain agents working for these companies or individuals (middlemen) who buy prawn from aqua farmers and sell it to the company.

The feed supplier arranges an agent from a marketing company to aqua farmers to sell their harvest. On the harvest day feed supplier comes along with the marketing agent, who weighs the prawn and after grading the count is decided to determine the price. The price of the prawn, as discussed earlier, is inversely proportional to the count, i.e., the lesser the count the higher is the price. The marketing agent picks up prawns randomly from crates and weighs them. The average of all these weights are considered as the prawn count. Later, the marketing agent gives a slip[4], which is the evidence for the aqua farmer that he had sold the prawn to the particular agent and the money is paid to the feed distributor later on production of the slip the following day. The feed distributor returns the balance money to the aqua farmer after deducting the cost of feed.

The feed supplier[5] at Allavaram was a native of Godilanka who is now settled in Allavaram. He is a relative to many aqua farmers. This is because they had sufficient capital and they were the earliest to venture in to aquaculture. Rajulu have large tracts of agricultural land and coconut gardens in the region. Aqua farmers who are also largely from their caste rely up on them. In addition to the feed distributors, there are 8-10 marketing agents in Amalapuram. Except two all are Rajulu. Diagnostic laboratories, which are mostly owned by Rajulu, prefer to employ technicians from their caste only.

Inter-Caste Relations

The traditional payments for washer man, barber, carpenter and *sonthamanishi* are in operation in aquaculture and their relationships to a certain extent are tending to be contractual and impersonal. The services of barber now are purely on cash payments. The carpenter is now engaged for making the sluice gates, check trays, etc., essential for the aqua farming and paid in cash. The individuals working both as *sonthamanishi* and aqua worker receive the traditional payments.

The dependence of aqua farmer on the labour of aqua worker, who belong largely to Mala and Madiga castes, has increased the bargaining capacity of aqua workers. The traditional kinds of patron-client relationships, which are still noticeable in the arena of agriculture, to some extent, are not noticeable in case of aquaculture.

IX

Impact of Aquaculture

Aquaculture has its impact both on the natural as well as social environment. In the social environment it has created new opportunities for men and affected employment of women. On the environmental front aquaculture has some negative impact on land utilisation and agriculture. Aquaculture requires saline water for rearing prawn. At the time of harvest all the water in the pond has to be drained or pumped out into the irrigation canals and also into *Kousiki.* The location of most of the ponds in the lower part of the village near Linepet and Penumalapeta causes inconvenience to the residents of this area, mostly SCs. In addition to the foul smell that it emanates, this pumped out saline water had been polluting the drinking water tanks located in these areas by turning them into saline. As a result, the residents of this village have been facing acute drinking water problem. This had made the people living in these areas to depend more on one or two Rajulu families, who allow them to draw water from their wells, while water can not be accessed freely from others' wells.

Environmental Problems

The cultivation of prawn depends on a steady supply of seed from the sea. A number of hatcheries, which depend on the 'mother prawn', have increasing demand to cater to their supplier. This is also true with the wild varieties. The individual fishermen engaged in this catch also catch other species, which come in to their nets and are thrown on the shore. This has a long-term effect on the depletion of wild varieties in the aquatic ecosystem, which will affect the food chain and ultimately the macro system (Kori, 1996; Shaji, 1996; Baird, 1993; De Walt *et al.*, 1993 and Reddy, 1995: 1).

Due to unplanned and uncontrolled growth, aqua farms encroach up on agricultural lands and important estuary ecosystems like mangroves. The mangroves are important as they harbour a rich variety of faunas and are essential for the protection of the coastline (Shiva, 1994: 22-24; Kocherry, 1995). In their study on Thailand, Flaherty *et al.*, observed that, "The expansion of shrimp farming into rice fields along the coast of the southern peninsula during 1990s enabled industry officials to boast about the increase in national production, and claim that development pressure had been taken off the remaining mangrove forests.

Government and industry officials may have believed that few parties, particularly the international NGOs, would concern themselves with conversion of low yield rice paddies in the south to shrimp ponds" (Flaherty *et al.*, 1999: 2054-66).

One of the main problems associated with aquaculture is water pollution and salination of ground water, a rampant problem in most coastal villages. Several coastal villages have already been affected and were abandoned because of the ground water salination due to unsound farming practices. In prawn farming, seawater is pumped into the land into man-made ponds for cultivation. These ponds are not lined and as a result, this water seep downs and mixes with the fresh ground water, rendering them unfit for consumption. The soil also loses its fertility and becomes saline. This makes the lands unfit for cultivation in future. Also, there is large-scale withdrawal of groundwater resulting in its depletion (Raj and Dharmaraj, 1995; Baird, 1993; Issac, 1995: 6-13 and Prasada Rao, 1995: 7).

The organic load of the effluent is high because of the excreta and other dead and decay matter in the effluence. This high organic load increases the Biological Oxygen Demand (BOD) level and when let into the sea or fresh water bodies there is a depletion of dissolved oxygen (DO) which will affect the respiratory process of the aquatic organism (Algarswami, 1995: 27-30; Shaji, 1996: 93-5; Krishnan, 1995:6-10, 1996:38 and Sekhar, 1995: 88).

Ammonia, which is released as a by-product of the metabolic process of the shrimp, can get converted into nitrates which enter the food chains of people. This can cause a condition called 'metheamoglobinaemea', reducing oxygen carrying capacity in haemoglobin (Srinivas, 1995: 14; and Krishnamohan, 1995: 5).

Socio-Economic Problem

Most shrimp farms are located near the coast, the inversion lure the poor fisher folk to sell their land to set-up aqua farm. Though fishermen get an immediate economic benefit, which is short lived, they are getting displaced from their lands. Apart from this, there are other problems, such as reduction in agricultural land, exposure to new values and traditions that affect the rural poor, etc. Apart from this, common lands, grazing lands, etc., are encroached upon causing problem to the villagers

(Krishna, 1995: 4-6; Rubino *et al* 1990: 386-94; Ruddle, 1993: 20-41; Peterson, 1982: 21-30).

The drained water stagnates in the canals and it has become a breeding place for mosquitoes. This has aggravated the incidence of malaria, according to the health staff and 'Doctors' (RMPs) of the village, in Godilanka in general and in Linepet and Penumalapeta areas in particular.

The yields of agricultural lands adjacent to the aqua ponds have come down significantly due to seepage from aqua ponds. Earlier the average yield in these lands ranged between 30-32 bags of paddy per acre for every crop. This is now reduced to 20-25 bags of paddy. The affected farmers do not complain to revenue authorities because the aqua farmers in this village happen to be their kinsmen or relatives. However, there are instances of farmers complaining to the caste *peddamanushulu* regarding the problems faced by them due to seepage from aqua ponds.

R. Seetharama Raju is a small farmer with 3 acres of agricultural land. He had a complained to *peddamanushulu* that the saline water from the aqua pond of Prasad Raju seeped into his agricultural lands and this had affected his paddy crop. Seetharama Raju had threatened to file a complaint with the concerned official to stop aqua farming of Prasad Raju. The *peddamanushulu* felt that any complaint to officials would create problems to all aqua farmers. The *peddamanushulu* called both Prasad Raju and Seetharama Raju and discussed the problem with them. They advised Seetharama Raju to convert his land to pond or lease it out to any one or to Prasad Raju. Seetharama Raju turned this down. Following this, they advised Prasad Raju to pay compensation for the crop damage, at the rate of 3 bags of paddy for every crop, or dig a one-metre width seepage canal all along the pond so that the seepage water can be drained out. Prasad Raju had agreed to pay for each crop money or paddy.

There are also cases where some of the affected farmers had no alternative but to either lease out land or resort to aqua farming, though reluctantly, due to continual decline of yield. Thus, some converted their agricultural lands to aquaculture ponds. A few acres of coconut gardens have also been converted in a similar way into aqua ponds as their yields have fallen and in quite a few cases coconut trees died due to seepage of salt water into coconut gardens.

Status and Role of Women in Aquaculture

The activities involved in aquaculture are onerous and women are not preferred. The job demands a greater amount of agility and they have to get into ponds. They have to stay on the pond during nights during the advanced stages of crop. All these are considered as limitations for women. Hence, they are not engaged for such activities. Thus they could not find any considerable position in the aquaculture farming. Frequently women become substitutes to their fathers or husbands, who work as aqua-watchmen, to watch the pond while they have gone for their meal or on any urgent work.

Women in aquaculture are employed during harvest days of the crop season. Two to three women will be employed for one day for an acre. They wash and clean the prawn after which they pack them into a container allocated for this purpose. On the harvest day their wages are more than the wages paid for agricultural operations, about Rs. 40/- to 50/- (Table 4.12). Since the labour involved is more, the wages are also higher. On the harvest day the workers are provided tea, snacks and lunch in the afternoon. They are also given some quantity of shrimp, which grows along with tiger prawn in the pond.

Table 4.12: Working days for Women in Aqua Farming

Type of activity	Employment of women	Days in a crop season
Preparation of pond	Optional/ 2-3 days	2-3 days
Stocking	Optional/ one day	One day
Monitoring the pond	Now and then as the situation demands	
Harvest	One day	2-3 days

New Occupations

The new occupations that have come into existence after the emergence of aquaculture in Godilanka are related to aquaculture. They are seed business, selling agricultural lime and dolomite/ zeolite, loose diesel vending, and accessories, such as nylon threads, check-trays, nets, etc. The services of a diesel pump-set mechanic are essential for aqua farming, as the pump-sets are used extensively.

About ten members from Rajulu caste in Godilanka are involved in seed business. They have their seed tanks located in Godi all along the shore since seawater was available abundantly. The land for the seed tanks is leased-in by the seed businessmen for a period of fortnight, before the aquaculture season starts. Each seed tank is square/rectangular in shape and is roughly half a *kuncham* (approximately 25 sq. yards) in size. The contracted money has to be paid in advance. Many persons from different caste groups and villages are engaged in this enterprise in Godi.

The seed business, more or less, lasts for a month. They procure the seed according to their requirement, which depends on acreage under aquaculture in the locality and average density of stocking for each pond of their clients, from the hatcheries located in between Kakinada and Visakhapatnam. The hatcheries breed and store the seed (post-larval form) based on the salinity specifications suitable to the local area. The seed businessmen specify the salinity levels to the hatcheries in advance so that they can receive their requirement in time.

The aqua farmer, after a close examination, selects the seed and places order. The seed businessmen engage women in counting the seed. For counting the seed a small triangle measure is used and hundred seeds thus counted are transferred into a plastic bucket, later transferred into felt containers. Some individuals transport the seed stored in 'felt containers' through mopeds (a motorised two wheeler) and the seed businessman pays them. These mopeds are specifically altered for this purpose. Some of the persons involved in seed business either have their own aqua ponds or have close relatives who own ponds. After meeting their requirements they market the seed to others. The seed businessmen are usually younger persons (between 19 and 27 years) and mostly indulge in joint aqua farming.

In Godilanka there are about 230 diesel engine pump-sets used in aqua farming and two mechanics are engaged in their repairs. These pump-set engines are used either for drawing saline water (backwaters) from *Kousiki* into the culturing ponds or for drawing ground water, which is used to lower the salinity of the water in pond. It is also used in the advanced stages of the crop to re-circulate the water when the dissolved oxygen content falls down below the desired level, as an alternative for aeration of the pond.

The engines are used almost daily right from the preparation of pond till the final harvest of the crop. It runs for about 3-4 hours daily. Some of the aqua farmers have bought used motor engine pump-sets at a cheap price. These pump-set engines require frequent servicing. For this purpose the aqua farmers approached a Rajulu caste mechanic from Magam, which is 12 kilometres away from Godilanka, who was repairing their diesel pump-set engines for two years, and invited him to settle down in Godilanka. He was given a rent-free accommodation in the central part of Godilanka so that he will be available whenever required to repair their pump-sets. A woman relative of the mechanic from Magam was married to a resident in Godilanka in whose house he was provided the accommodation. Besides, he even had distant relatives in Godilanka, which also prompted him to settle down in Godilanka with his family in 1997. He earns Rs. 200 to 300 per day.

In addition to this Rajulu mechanic, there is a Mala caste man from Pathuru, who had experience in foundry work, involved in repairing diesel pump-set engines. He attends to repairs mostly in the adjacent villages and, occasionally, aqua farmers from Godilanka approach him, depending upon the urgency. His earnings from repairs vary between Rs.150 to 200 in a day.

In Godilanka, there are two loose diesel vendors from Rajulu caste, who have their shops near Gogulamma temple and Primary Health Centre. They sell diesel at an additional charge of Re. 1/- per litre and bulk purchasers get some rebate. They give diesel on credit to reliable aqua farmers as they clear their dues periodically. The shop near the PHC also sells agricultural lime, dolomite, and, depending on the timing of the crop, sells small quantities of feed and fertiliser also. The loose diesel vendor at Gogulamma temple sells accessories like nylon threads, check-trays and nets necessary for this enterprise.

By and large, feed, in general, is bought from the dealers belonging to different companies located at Allavaram and Amalapuram. A government employee working in Amalapuram sells agricultural lime and dolomite. He stores them in his storehouse located near the High school. An aqua farmer maintains a little higher quantities of aqua feed, more than necessary for his use, and sells them in small quantities to others in times of need.

NOTES

1. Buy–back policy was introduced in 1991-92 by feed companies both national and international, viz., Water Base Luxe, CP (Charoen Pokphand), Higashi, NASA, DCL Maritech, Avanti, ITC, Godrej, Somken Feeds and others.

2. A mechanical device used for digging and removing earth. This they viewed to be economical and faster when compared to manual digging, which requires large manpower.

3. The water quality parameters according to MPDEA (1991) is as follows:

Parameter	**Range**
Temperature	26 – 33° C
Salinity	10 – 35 ppt*
Transparency	20 – 35 cm
Dissolved oxygen	3 –10 ppm**
pH	7.5 - 8.5
Total Ammonia	less than 1.0 ppm
Free Ammonia	less than 0.25 ppm
Nitrite	less than 0.25 ppm
Hydrogen sulphide	less than 0.25 ppm
Heavy metals	
Mercury	less than 0.0025 ppm
Copper	less than 0.1 ppm
Iron	less than 0.01 ppm
Zinc	less than 0.25 ppm
Cadmium	less than 0.15 ppm

* ppt - parts per thousand

** ppm - parts per million

4. A small white paper, which has the details of weight of prawn count and the corresponding price of count on that day and the signature of the agent without any stamp of the company.

5. Most of the feed suppliers (80%) in Amalapuram are Rajulu. About15 feed supply companies and about 25 distributors are located in Amalapuram. Outs of them 20 belong to Rajulu and the rest belong to Vysya (1), Kapu (2) and (1) and others (1).

Chapter 5

Socio-Economic Development and Changing Agrarian Relations

There is hardly any village economy today where economic growth has been internally generated. In a rural community the economic growth is, in general, the result of an interaction between external stimulus and internal response. New opportunities for villages are either the result of conscious intervention by the State or other external agencies or may be a consequence of a private initiative. The village as a whole or a part of it may accept change, which depends on the attitude, values and response of the recipient community. Thus, it is necessary to understand the entire arena of new opportunities that have emerged in the villages as a result of adoption of new innovations, technology, etc., and analyse their implications on the socio-economic life of the rural community at large. Godilanka provides a good opportunity to analyse the impact of capital intensive aquaculture over the labour intensive agriculture. The process involved in the conversion of agricultural lands to aqua farms due to the introduction of aquaculture and its consequences provide an opportunity to delineate the changes that have occurred in Godilanka.

This chapter deals with the socio-economic and cultural changes that have occurred after the introduction of aquaculture in Godilanka. This encompasses different aspects that are observed as part of a village social system, viz., economic, social, cultural and political. The existing institutions had altered and adapted to the changing conditions/situations.

The land utilisation pattern in Godilanka has changed due to some change in the production system with the introduction of aquaculture.

This has resulted in the adaptation of farmers, cultivators and agricultural labour to the changed circumstances. Monetary consideration involved in the production of prawn has altered the perception of landowners, and tenants as well, towards agriculture and coconut farming.

The introduction of aquaculture has brought in to Godilanka a new type of tenancy in agriculture, *gutta vyavasayam*. This had contributed partly in altering the traditional relations between landowners and labourers. The categories of attached labour, such as *Paleru* and *sonthamanishi*, have switched over to newly emerged occupation of aqua workers. Aquaculture required greater surveillance of the aqua ponds. Because of this, the landowners continuously needed the presence of labour (aqua workers) on these ponds. This had led to greater dependence of aqua farmers on labour, which has consequently increased the bargaining power of labour, resulting in higher wages for aqua workers and others.

After aquaculture in Godilanka, organised labour groups, *bantapanivallu*, have entered into Godilanka from the neighbouring villages. They could be contracted *en masse*, who also move to other villages in the region. This *bantapanivallu* would work to dig new aqua ponds, cleaning the debris in aqua ponds and also in agricultural operations, such as transplantation and harvest. Earlier, *sonthamanishi* used to arrange labour for his landowner during agricultural operations. Since many *sonthamanishis* have become aqua workers, the landowners have to arrange labour by themselves for peak agricultural operations. Under these circumstances, they find it convenient to engage *bantapanivallu* for agricultural operations also. This facilitates the cultivators to complete the work faster and also economically, as the need for maintaining attached labour becomes unnecessary. The shift from agriculture to aquaculture also resulted in a reduction in the participation of women in workforce, as their participation in aquaculture is minimal.

The income of aquaculture had enhanced the living standards of the aqua farmers. They have invested their profits in buying agricultural land, coconut gardens, additional aqua ponds, etc., both within and outside the village.

Aquaculture had its impact on the caste relations and caste *Sanghams*. This no way implies that aquaculture alone was responsible for the consolidation of caste *Sanghams*. Many other factors did play a significant role in this process of consolidation. This had its repercussions among different caste groups and their interrelations, both within and outside the village.

I

Godilanka is basically an agrarian economy. Introduction of aquaculture in 1990s brought in dramatic changes in Godilanka's economy, in particular, and its agrarian relations, in general. The process of agriculture requires investment for procurement of seed, fertiliser, pesticides, etc., and labour for nursery transplantation, weeding, harvest etc. About Rs. 4,825/- per acre is needed as investment for a crop season. The average yield is about 30 bags of paddy per acre. In 1998, each bag of paddy was about Rs. 310/- and the returns were Rs. 4,475/- per acre after the deduction of costs of inputs.

Table 5.1: Investments and Labour in Agriculture (in Rs.)

Type of Activity	Investment	Labour
Dukki	-	400
Aku (nursery and fertiliser application)	260+140	150
Transplantation	-	675
Weeding	-	400
Fertiliser application	800	50
Pesticide	1300	50
Harvest (tractor)	600	500
Miscellaneous	-	800
Total	3100	3025

On the contrary, aquaculture involves a higher investment while the labour employed is to the minimum. The investment is needed in two stages, first, at the time of development of land into a pond and later at the time of aqua farming. The first stage involves conversion of land, installation of tube-well (to tap underground water) and a diesel pump set; and accessories such as feeder canal pipes, sluice gates, dykes. Approximately at this stage Rs. 50,000-60,000/- are required for a pond. The extent of pond or water spread area (WSA) may vary form one acre to a hectare. First stage ends with the establishment of pond and simultaneously the second stage of raising a crop begins. During this stage, fertiliser and manure for pond preparation, procurement of seed and feed, diesel for pump sets and miscellaneous things like dolomite/ agricultural lime are required. Beside, aqua worker for the crop season, Markalu and labourers on the harvest days will be employed. The expenditure incurred by an aqua farmer for raising a crop per acre in a season will be nearly one lakh rupees. The type of farming practised in Godilanka is semi-intensive and produces a yield of about 500-600 kgs. of prawn. The market prices fluctuate and during 1998, 30 count of catch was about Rs. 400/-.

During the first crop of a pond after deducting the infrastructural costs, the profit margin will be very low and a farmer would receive will be about 50,000 per acre. From the second crop onwards there will be recurring profit for every crop and it will be more than one lakh rupees per acre. As discussed in Chapter 4, one has to bear in mind, that a major portion of the operation costs incurred is towards feed, which will be supplied by feed distributor on credit. In some cases, even the diesel pump-set is procured against credit on a pre-condition that it will be cleared with interest after the first crop harvest.

Table 5.2: Investment and Labour in Aquaculture

Development of the Pond Activity	Investment	Labour
Earthen work (digging)	-	15,000-20,000
Underground Bore	10,000	
Pump-set and accessories	20,000	
Operational costs		
Fertiliser and Organic Manure	600	Aqua worker
Seed	6000-9000	@1200 for 5
Feed	70,000-80,000	months 6,000
Diesel	5000	
Miscellaneous	5000	
Total		21,000-26,000

In both agriculture and aqua farming labour is an essential component. Agriculture is primarily labour intensive and both men and women are employed. Certain agricultural activities are allotted mainly to women such as transplantation, weeding and harvesting. Men exclusively practise some of the operations such as *dukki*, application of fertiliser and pesticide, etc. Rest of the operations involves both men and women (refer Table 3.13). Landowners employ attached labourers to work with them. In addition to them, 'private' labour are continuously hired in almost all of the agricultural operations.

In aqua farming, usually, the pond is developed just before the commencement of the crop season. Labour required in aqua farming is minimal. An aqua worker is hired to look after the pond from the stage of seed to harvest of the prawn and he can take care of aqua pond up to 2.5 acres. Additional labour is engaged in the preparation of pond, embankments, repairing of canal and during harvest days. Women are engaged on the days of harvest only. Markalu from neighbouring villages are engaged after two months of feeding the seed, once a week or ten days, to monitor the growth of the prawn by the pond owner. This type of activity is continued until the harvest. Three Markalu and three women are usually employed on the harvest day.

Table 5.3: Additional workdays in Aqua Farming

Type of Activity	Men	Women
Preparation of the Pond	1	-
Embankment	2	-
Repairing inlets and Canals	2	-
Harvest	3	3
Total	8	3

*Markalu employed are from neighbouring villages.

Cropping Pattern

In Godilanka, Paddy is the staple crop and along with coconut it is primarily produced for market. They produce paddy for Kharif and Rabi, and third crop raised is either *rotta* or pulses. This pattern continues to be the same even after the emergence of aqua farming. With the emergence of aquaculture the land under cultivation has reduced drastically, from 505 acres to 351 acres, a decline of 30.5%. In 1993 owners cultivated about 328 acres (65%) of land by engaging hired labourers. Presently, in 1998, it is reduced to 129 acres (36.8%).

Prior to aqua farming, land under *paalu vyavasayam* was the only form of tenancy with176.75 acre (35%). In 1998 land under tenants has increased to 222 acres (63.2%) with 121 acres (34.4%) under *paalu vyavasayam* and 101 acres (28.8%) under *gutta vyavasayam*. This increase in land under tenant cultivation was due to a decline in the cultivation by owner cultivators.

Table 5.4: Land under Different Types of Cultivation

Agriculture Type	Earlier	Now
Own	328.25 (65)	129 (36.8)
Paalu vyavasayam	176.75 (35)	121(34.4)
Gutta vyavasayam	-	101(28.8)
Total	505 (100)	351(100)

Note: Figures in parentheses are percentages.

Prior to the introduction of aquaculture in Godilanka, there were 110 landowner cultivators cultivating paddy. After the introduction of aquaculture the proportion of landowner cultivators had significantly declined (42.7%). The entire decline can be observed only among the Rajulu landowners. Among the cultivating landowners (96), a significant number of them have taken to aquaculture (47). Among the others who owned land, Settibalija (8) and Mala (6) still continue to practise agriculture (Table 5.5).

Table 5.5: Landowners Practising Agriculture by Caste

Caste	Rajulu	Settibalija	Mala	Total
Prior to Aqua	96 (87.3)	8 (7.3)	6 (5.4)	110 (100)
After Aqua	49 (77.8)	8 12.7)	6 (9.5)	63 (100)

Note: Figures in parentheses are percentages.

Tenants

Prior to the introduction of aquaculture, as discussed earlier in chapter 3, *paalu vyavasayam* was granted to *sonthamanishi* or his family members.

In *paalu vyavasayam*, the harvest is shared on 1:2 for *tholakari* and 50:50 basis for *sarava*, after deducting investments. The agricultural inputs such as seeds, fertiliser etc., cost Rs. 2,500 per acre and after deducting this, *paalu* farmer will be left with 7 bags of paddy (Rs. 2266/-) per acre as part of his share. Out of this, about three bags of paddy go towards costs of labour. Ultimately, he is left with four bags of paddy, during *tholakari*, which he utilities it for his own consumption. In a similar fashion, in *sarava* after deduction of capital inputs and labour charges he gets his share of 7 bags of paddy. Thus, totally a *paalu* farmer gets an annual share of 11 bags paddy. Contrary to *paalu*, in *gutta vyavasayam* the tenant is more independent and has to invest the in the costs of cultivation. A *gutta* farmer gets 14 bags of paddy annually.

The form and type of tenancy and the mode of payment involved varies between agriculture and aquaculture. In the agriculture situation the only form of tenancy was *paalu vyavasayam*. Land was granted for *paalu vyavasayam* to *sonthamanishi* (attached labour). In some cases, it may be family members or relatives of *sonthamanishi* who work for the landowner as *Paleru* looking after the livestock and attending to the household needs of the landowner. The senior most *sonthamanishi* acts as supervisory *sonthamanishi* for the landowner's un-leased lands kept for own cultivation. Supervisory *sonthamanishis* were observed largely from Settibalija caste. There were also some supervisory *sonthamanishis* from Rajulu, Settibalija and Mala castes. These supervisory *sonthamanishis* arrange for labour and look after the requirements of paddy fields, coconut gardens, etc., of their landowner. *Paleru* and *sonthamanishi*, by and large, belonged to the Mala and Madiga caste groups. In *paalu vyavasayam* entire agricultural inputs, viz., seeds, fertiliser, etc., would be supplied by the landowner. The sharecropper has to procure labour necessary for the piece of land given to him for *paalu*.

After the introduction of aqua farming in Godilanka, a new type of tenancy, *gutta vyavasayam,* has also come into existence. This coexists with *paalu vyavasayam*. In *gutta vyavasayam* all the necessary inputs and labour have to be taken care of by the tenant. A fixed amount of kind, usually grain payment, has to be paid to the landowner after the harvest. Here the by-product's, such as straw, would also be given due importance. The control and distribution of straw depends on mutually agreed conditions at the time of the contract. In *gutta* the landowner is not concerned about the fate of the crop. Irrespective of the conditions of crop, the tenant has to pay the agreed amount of grain to the landowner. One has to understand that in *gutta* the role and indulgence of landowner is limited.

There is some kind of polarisation that has been taking place in Godilanka. While Rajulu and Settibalija are resorting to *gutta*, Mala and Madiga continue to be under *paalu vyavasayam*. The landowners prefer giving their lands on *gutta* to the *sonthamanishis*, who had worked with them earlier.

With the introduction of aquaculture, a shift in production has taken place. However, this shift has not affected the ownership of the means of production. While the owner cultivators among Rajulu have become supervisory landowners, the Settibalija *sonthamanishis* have now become *gutta* tenants. In line with this, some from the Mala as well as Madiga castes have taken to *paalu vyavasayam*. Earlier most of them were agricultural labourers. Thus there is some change in the agrarian social structure of Godilanka with regard to the kind of people involved or engaged in agriculture.

Wages

The conditions of labour—such as wages, days of work, etc.—prior to the emergence of aquaculture show some significant differences with those after the emergence of aquaculture. In case of attached labour, while the nature of work remained same all round the year, their wage rates have gone up. On the other hand, the number of days of work for 'private' workers has come down while their wage rates have gone up. Aqua workers being a new addition, most of the earlier *paleru* and *sonthamanishi* have shifted to this new occupation. Now, they work for ten months in a year as aqua worker and the remaining period they work as 'private' labour (Table 5.6).

Occupations

The shift from agriculture to a combination of system of agriculture and aquaculture has altered the occupation structure and a new category of aqua farmers and aqua workers come into existence. In the present scenario of agriculture and aquaculture, some occupations have been retained in agriculture and a few have geared to the new context.

Table 5.6: Status of Labour before and after Aquaculture

Agricultural Labour	Days of work	Wage rate (in Rs.)	Annual wages (Rs.)
Prior to 1993			
Paleru	Entire year	20 bags of paddy annually	5600/-
Sonthamanishi	Entire year	*Kunchudu vodlu* (20/- appr.)	7200/-
Private	20-22 days a month	30/- per day	8000/-
In 1998			
Paleru	Entire year	24 bags of paddy	7920/-
Sonthamanishi	Entire year	5 *manelu* + 2-*beta* (30/- appr.)	10,950/-
Aqua worker	10 months a year	Monthly 1200-1500/-	9 - 12,000/-
'Private'	15-20 days a month	50/- daily	12 -15,000/-

The agrarian situation has supported a system of occupations, viz., farmers—who either leased-out or did self-cultivation by employing labourers, tenants (*paalu* farmers) and labourers, (*paleru*, *sonthamanishi* and 'private'). There were a few *dimpu* workers, belonging to Madiga caste, engaged in coconut gardens for harvesting coconuts because of their skill and expertise. The *paleru* and *sonthamanishi* were contracted on an annual basis and given an interest free loan at the time of contract. This amount will be reclaimed gradually in their wages (Table 5.8).

Presently, fewer landowners are engaged in agriculture and there is a tendency among the landowners to switch over to aquaculture. There are also some landowners who are both agriculturists and aquafarmers. Among tenants, both *paalu* and *gutta vyavasayam* is practised. *Gutta vyavasayam* is taken up by individuals as well as jointly by two or more persons on sharing basis. In the category of labourers, there is *paleru*, *sonthamanishi*, aqua worker and 'private' workers. Except for aqua workers, all others exist in agricultural situation. Aqua worker is a labourer engaged on aqua farms. He is usually contracted for a crop season. However, there is a tendency among the aqua workers to continue with the same aqua farmer. They receive an interest free consumption loan, more or less, on par with *sonthamanishis*.

Table 5.7: Labour Comparison by Type

Type of Labour	Earlier	Now
Paleru	36(16.1)	16(6.5)
Sonthamanishi	92(41.1)	56(22.7)
Sonthamanishi-cum-Aqua worker	-	21(8.5)
Aqua worker	-	43 (17.4)
'Private'	96(42.8)	83+28* (44.9)
Total	224(100)	247(100)

* Tied for a crop season.
Figures in parentheses are percentages.

Aqua worker will be out of job for about two months in a year. Earlier *paleru*, *sonthamanishi* or 'private' labour had reoriented and adjusted themselves to aqua farming due to higher wages. In the category of *sonthamanishi*/aqua worker, there are some who work as *sonthamanishi* during agriculture season and on the other days work as aqua worker.

Work, Working Pattern and Wages

In agriculture, labour employed were attached, viz., *paleru*, *sonthamanishi* and 'private'. The attached labourers work on annual basis while the 'private' works on daily wages. These labourers either work on paddy fields, coconut gardens or household and other associated works. *Paleru* works from dawn to dusk daily, about 12 hours, whereas *sonthamanishi* works for about 9 to 10 hours in a day and 'private' agricultural labour works for about 8 hours a day. The wages of *paleru* and *sonthamanishi* are settled once in a year and the accounts of a *sonthamanishi* are monitored every month by the landowner. The attached labourers are advanced an interest free loan at the time of entering into the contract, which are renewed every year. The 'private' labour is paid wages in cash on the same day. However, on the harvest days 'private' labour can take their wages in kind. In addition to wages and loan, attached labour receive traditional payments in the form of two pairs of clothes, food and cash gifts during festivals. In addition to this, the supervisory *sonthamanishi* also receives some grain payment after every crop harvest. Mostly, the payments for these categories are in cash equivalents of kind. Earlier, they were paid in kind. The *dimpu* persons are paid once a month and they also receive consumption loan and a part of wages would be retained against the loan.

In aquaculture the aqua worker is paid a monthly wage, which is higher than that of *sonthamanishi*. He receives consumption loan, which is lesser than that given to *sonthamanishi*. *Sonthamanishi*-aqua worker is paid wages similar *sonthamanishi* and also receives customary payments.

Women Labour

Earlier women were employed both in agriculture and in coconut gardens during harvest days. They were employed in all agriculture operations. They were preferred in agricultural activities like transplantation, weeding and harvest. These agricultural operations are more arduous than those undertaken by men in agriculture. On these days women would be paid *kunchedu vodlu* in kind by large and big farmers, and its equivalent in cash by other cultivators (see chapter 3).

It is obligatory for women of attached labour to work in the fields of the landowner during peak agricultural season and during this period they are paid wages less than the others. Women have to work for about 2 months in each crop season, which has now been reduced to half this period. As a result, many Scheduled Caste women now go for work to neighbouring villages during peak agricultural season. However, the magnitude of work was not unaffected in coconut gardens.

The switching over to aquaculture favoured men to the disadvantage of women, as aquaculture needed more agility, operation of mechanical pump sets, feeding prawn during night times, to get into pond for treating pond with dolomite fortnightly and run other errands, which were difficult for women to cope with. However, on the harvest day about 2-3 women would be engaged for an acre of pond to wash and clean the harvested prawn. Nevertheless, the quantum of women labour displaced due to conversion of agricultural land to aqua farms could not be accommodated in aquaculture or agriculture.

Now these displaced women are working as house or servant maids in landowner's houses to supplement their family incomes. Some women from landless Rajulu, Settibalija, Mala and Madiga families have taken up to dairying and maintaining milch cattle. A few women, especially from the Scheduled Castes, started migrating to Gulf. These women send remittances, which are used to improve their standard of living.

Though wage rates have gone up, their workload is now reduced considerably to about one fourth. A woman's role has changed from one of the main contributor to family income to a secondary contributor to family income. Thus the structural role of women has been transformed from relatively independent to totally dependent, creating new types of bondage in the family. Their position has now become more vulnerable than before.

Displacement of women from agricultural work has given rise to more free time than before and this has increased their interaction among themselves leading to some organised activities. At Linepeta recently a Mahila Mandali was established and it had sought assistance from Government for getting sewing machines under TRYSEM programme.

Levels of Living

The income pattern in Godilanka varies and it is dependent on the ownership of land or aqua farm. It determines the expenditure and living standards of a household. It is observed in the earlier chapter that income in aqua farming is much higher than in agriculture and thus, many landowners have switched over to aquaculture. A considerable number of tenants are now engaged in *gutta vyavasayam* and their income levels have gone up due to increase in returns.

The life-style of a big landowner can be understood from the case study of Satyanarayana Raju, a 68 years old Mandal Parishad Territorial Committee (MPTC) member. He possessed 20 acres of paddy land and 16 acres of coconut gardens. He used to maintain two *palerus* and six *sonthamanishis* to work on his fields. Satyanarayana Raju employed a supervisory *sonthamanishi* from his caste to look after his fields. After harvest of every crop, some grain was kept for consumption and the rest was marketed. He used to get income both from agriculture as well as from coconut gardens. He has been politically active and maintains relations with leaders in the local area. He leads a political party in the village.

He has two sons and two daughters, who are all married. Prior to aqua farming, both his sons (graduates) were living in Vishakapatnam doing their own business. Every year during summer their families come and stay in Godilanka to spend time with their parents. His second son has returned to Godilanka along with his family to undertake aquaculture and has been staying with his father since 1994.

After introduction of aquaculture, 9 acres of land was converted into 6 ponds and in 1997 he leased out 5 acres and his second son is cultivating the rest. Now, a *paleru* and 3 *sonthamanishis* are working on his fields and 2 *sonthamanishis* became aqua workers working with his son. Satyanarayana Raju is now detached from agriculture but he visits fields frequently. His son is supervising all the activities and he assisted by the supervisory *sonthamanishi.* All the accounts relating to paddy and coconuts are still maintained by Satyanarayana Raju. With the addition of his son's family, the household consumption has increased. The frequency of consumption of non-vegetarian has increased.

From the earnings of aquaculture, Satyanarayana Raju had purchased more coconut gardens in the neighbouring villages, as the prices of coconut gardens were lower there than in Godilanka. In 1998, he had changed the roof of his ancestral house from tilled to RCC. His son had installed individual dish antenna for viewing satellite television channels in 1998, as owning of it has become a matter of prestige in the village. Satyanarayana Raju now lends money to landowners of his community. In 1997 he has advanced approximately 80,000/- and in 1998 about 98,000/- as credit.

Like Satyanarayana Raju's son, about 12 individuals had returned to the village after the introduction of aquaculture. All of them are now involved in aqua farming. In 1996 Satyanarayana Raju's son has brought a scooter and he frequently visits Amalapuram for necessary work pertaining to aquaculture and for other purposes. The frequency of playing cards has come down as most of the time he spends at the pond during the crop season but during the occasions of social functions they playing cards with higher stakes. He admitted his son in a residential school at Amalapuram.

The case of Naga Srinivas gives an understanding of the life-pattern of a typical landless labour in the present situation. Naga Srinivas is in his thirties. Prior to 1993, he was living along with his parents and an unmarried brother and all the members were contributing to the family income. They possessed half-acre of coconut garden. His father worked as *sonthamanishi* and he continues to work with the same landowner as *sonthamanishi*. Throughout the year both of them have work. Women members of the family (mother and wife) will be busy during the crop season and intermittently would be working in coconut gardens and attending other works. The incomes they earned could hardly support in maintaining their family.

After the coming of aquaculture, their wages had increased and they are continuing as *sonthamanishi* with the same landowner. Families living at the lowest level of income, like that of Srinivas, give primacy to food to ensure their survival. Srinivas complains that even with increased wages they could not live comfortably. Yet, he appreciates his landowner's generosity in advancing him money to keep his family going. Subba Raju, his landowner, explained that Srinivas's father has to arrange him labour whenever he needs. Thus the need for assured continuous labour on his fields is met. Moreover, there exists a relationship (he still remembers Srinivas's grand father who worked with his father), and this cannot be broken easily by either party.

At present, his *santa* (market) expenses are between Rs.75/- to 100/- in a week where he procures rice, pulses and other essential commodities. The frequency of consuming non-vegetarian has increased to twice a week, which was earlier once in 10 days. He is attending *Sangham* meetings regularly in which he had little interest earlier. In 1997, in addition to the clothes given to him as part of traditional payments, he brought clothes from Amalapuram for his wife and children worth Rs. 600/-. He purchased an old bicycle for Rs. 500/-. In 1998, he purchased a buffalo for Rs. 5,000/- borrowing from his master, which is looked after by his wife. She sells milk and this income is set apart to clear the debt made for the same.

Credit Structure

In Godilanka, it was customary for the landowners to advance interest free consumption loan to attached labourers working under them as *paleru* or *sonthamanishi*. As discussed earlier (Chapter 3) some agricultural labour families have been working with landowners families for a long period or even for generations. So it was essential for the landowners to retain them for the continuation of labour in their agricultural fields or coconut gardens. In order to retain and continue these labour the landowners extend consumption loan for the maintenance of labour working under them. Consumption loan was a security for the landowners to ensure labour supply during peak agricultural seasons. With the increase in demand for male labour after aqua farming, their bargaining capacity has increased. This has resulted in advancing more credit to these attached labour categories.

The 'private' labourers borrow money from moneylenders, who are also landowners. Here money is given after executing proper documents and after pledging their property (either agricultural land or house). The document contains the details of loan amount, rate of interest, which is usually Rs. 3/- to Rs. 5/- per month for every Rs. 100/-. The relationship between moneylender and 'private' labour is tending to be more impersonal than was the case prior to aquaculture. There is a new trend observed among the Scheduled Caste women to work as house/servant maids or as nurses in the Gulf countries for which they borrow huge sums of money to pay to the recruiting agent.

Engaging Labour from outside

Prior to the emergence of aquaculture there was no involvement of outside labour in the village. The landowner-labourers were in a state of dynamic equilibrium. The introduction of aquaculture in Godilanka has brought various groups into the village production system.

Large farmers converted their fields into aqua farms for which they had engaged a group of Uppara[1] caste members (about 60 members, men, women and children) from the uplands of Rangapuram village of Ramachandrapuram Mandal of East Godavari district in 1995. They had camped in Godilanka for about three months from December to Ugadi[2] (March). They came along with their families and stayed in the coconut gardens. During their stay in Godilanka they also worked in the neighbouring villages. Again they had visited Godilanka in 1997 (November-April) as they were engaged by a few landowners.

Banta group was engaged for the first time in Godilanka by a Rajulu landowner. *Bantapani* labourers, discussed in earlier chapters, are engaged both in agriculture and aquaculture activities. In agriculture, they are mostly engaged in transplantation and harvesting. In aquaculture, they are more often engaged in cleaning the pond bottom and also in converting agricultural land to aqua pond. They cleaned about 25 acres of aquaculture pond in 1997 and 40 acres in 1998. There are about 3 *bantapani* groups that frequent Godilanka.

Disposal of Cattle

Prior to the emergence of aquaculture, farmers maintained cattle, either plough and/or milch animals. Plough bullocks were used for ploughing both agricultural fields as well as coconut gardens. Most of the farmers used to engage *paleru*, who took care of these cattle. They were paid annually and were given daily meals, etc. They also received customary payments and annual gifts during festivals, and consumption loan, which was interest free. The fodder requirements were met from the by-products of agriculture (straw) and green grass that grows on the bunds of the paddy fields. This was the rhythm of cattle, fodder and *paleru*, which was part of Godilanka for years.

The conversion of large areas of agricultural fields to aqua farms has disrupted the existing balance. The effective land under agriculture was reduced and with the changes in tenancy from *paalu* to *gutta vyavasayam*, a tenant is not obligated to share paddy straw with the landowners. The saline water in the aqua ponds had limited the growth of green grass on the bunds, which created a shortage of fodder. In addition to this, an increase in the wages for *paleru* and their maintenance has become burdensome for the landowners. This had also turned out to be uneconomical for many medium, small and marginal farmers. In view of these developments, most of the small and marginal farmers disposed off their cattle. In 1995 about 20-25 landowners from these categories sold their milch cattle and plough bullocks. Medium and large farmers have also reduced the number of cattle they maintained earlier. The small and marginal farmers now employ tractor or power tiller(s), which are hired, for tilling lands. They plough coconut garden by hiring plough bullocks on a rental basis. A few landless and marginal farmers from Mala caste now maintain plough and bullocks to meet the demand (Table 5.9).

Table 5.9: Persons Maintaining Cattle by Caste

Caste	Milching Cattle	Plough Bullocks
Rajulu	5	2
Kapu	1	-
Settibalija	4	4
Mala	8	5
Madiga	2	-
Total	20	11

Hut Hotels

Prior to aquaculture, there were two hotels maintained by Settibalija women and men. A new hut hotel (referred to locally as Kapu hotel) was established in 1996 by a migrant worker who had come to Godilanka the proceeding year to work as aqua worker. Owing to his ill health he could not continue as aqua worker. He established a hut hotel in the lower part of the village near PHC that caters to the needs of those involved in aqua farming, like snacks, tea and meals. Most of the people visiting this hotel are from Linepet, Penumalapeta, and Ballavaripeta (Mala and Madiga caste groups). A few aqua farmers, though but do not dine in the hotel they prefer to eat the food from this hotel at the loose-diesel vending shop, which is located on the other side of the road, opposite to this hotel.

The consumption of snacks for breakfast has increased in all the hut hotels. On an average, the earnings for all these hotels have increased after aquaculture. For instance, prior to aquaculture, the hut hotel at high school approximately used to get an income of Rs. 60/- to Rs. 100/ - per day against Rs. 100/- Rs. 175/- today. The consumption of tea has increased multifold, about 100-150 cups of tea is sold everyday much tea is also consumed by passers-by. Except Kapu hotel none prepare meals. They prepare meals against order and this has been helpful for visitors, officials, technicians and aqua workers during harvest days.

Land Value

The value of lands either paddy or coconut garden depended on the productivity of land/yield. The agricultural fields are distinguished as *saudu* and *gandera*. The production on *gandera polamu* is higher than that of *saudu polamu*. Due to the variation in production, the land rates are higher for *gandera polamu* over *saudu*. Prior to aquaculture, the land values varied between one to one and half lakh rupees per acre for *gandera polamu* and forty to fifty thousand rupees per acre for *saudu polamu*. The price of coconut garden was approximately 2.5 lakh per acre.

After aquaculture the entire scenario has changed. The land values have increased substantially to a notable level. In 1998 the cost of *saudu polamu* has gone up to one lakh to one and a half lakh rupees per acre, whereas the cost of *gandera polamu* has gone up to two lakhs to two and a half lakh rupees per acre. Correspondingly, there was also an increase in the value of coconut gardens, and they range from three lakhs to three and a half lakh rupees per acre.

It is pertinent to note that the rate of growth is higher in case of *saudu polamu* (2-3 times) and lower for coconut garden (about 50%). The location of *saudu* lands nearer to *Kousiki* and their suitability for aqua farming has increased their value. The high profits involved in aqua farming brought more money into circulation. While there were only a few to sell their land, there were many to buy, even at high prices. To a large extent, the purchasing capacity of aqua farmers has influenced the prices of lands.

The individuals who have disposed their lands were employers (working at distant places) and businessmen who settled in urban places, for whom it was difficult to look after their lands. With profits in aqua farming, many intended to invest on land. In Godilanka land being scarce, farmers purchased lands from neighbouring villages. Small and marginal farmers and landless agricultural labourers/tenants continue to invest in agricultural lands, while large and big farmers procured large tracts of land for both agriculture and coconut gardens in the neighbouring villages. The role of kinsmen and relatives is quite important in these transactions, especially among landless, marginal and large farmers (Table 5.10).

Older and Younger Generations

In Godilanka, traditionally all categories of Rajulu farmers engaged agricultural labourers (attached, casual) in their lands or coconut gardens. While the older members in the family were controlling agricultural activities, the younger members assisted them by overseeing, supervising and monitoring all the activities.

Table 5.10: Acquiring of Agricultural Land by Different Categories

Type	Extent in acres	Place of Buying
Land Less	14	Father-in-law's or distant relatives
Marginal Farmer	8	Distant relatives
Small Farmer	2	Neighbouring villages
Medium Farmer	15	Within village
Large Farmer	60	Distant relatives
Big Farmer	-	

By and large, among Rajulu the middle aged and youth are involved in aquaculture. The feed supplying companies encouraged these people by providing information and educating them. After analysing the lucrative business of aquaculture the middle aged and the youth among the landowners have adopted aquaculture. The buy-back policy had set the pace for aqua farming. In the households where both agriculture and aqua farming is practised, it is observed that elderly members take care of agriculture and coconut garden, while the younger members take care of aqua farming. In case of Settibalija both the younger and older members are still engaged in agriculture only.

A similar trend that is observed among Rajulu can also be noticed among Mala and Madiga, who constitute largely the labour either in agriculture or aquaculture. The aqua workers engaged on ponds belong largely to younger generation as the work demands more physical agility and involves operation of mechanical devices, staying on the pond during night, etc. The older generation is by and large continuing to work as attached labourers, *sonthamanishi* and as *paleru.*

Viewing Economic Trends/Production Strategies

Earlier to aquaculture in Godilanka, land was put to use to raise paddy and coconuts on a commercial basis. Agriculture was primarily meeting the subsistence requirements of the villagers and after this it was marketed. On the contrary, coconut was produced for market only. A larger section of population depended on agriculture compared to coconut gardens.

Coconut gardens require minimal labour. For instance, an acre of coconut garden needs 2-4 coconut harvesters, who are skilled and 2-3 labourers for collection and transporting coconuts to godown. The gardens are ploughed three times annually. Thus in view of labour employed and risk by natural disasters is less for coconut gardens. Farmers considered coconut gardens as higher to agriculture and it was more acceptable to medium, large and big farmers. However, agriculture was of importance to the landless labourers, small and marginal farmers.

After aquaculture the lands near *Kousiki*, which were once considered inferior, became best suitable for aquaculture. Though aquaculture involved with high risk, it was preferred due to higher margins of profit. Success of a few relatives in the village and in the region catapulated the others into aqua farming. The margins of profits involved overshadowed the risk factors. Aquafarmers view that one good crop can earn them money equivalent to what they would otherwise earn in ten years of agriculture. At present, highest preference is given for aqua farming, followed by coconut gardens and agriculture by the medium, large and big landowners who have lands all along *Kousiki*. However, paddy cultivation has its own importance for tenants, small and marginal cultivators and labourers. Thus, aquaculture sharpened the distinction between the big, large and medium farmers on one side and the small and marginal on the other.

Marriage

It is important to note that aquaculture had its bearing on the marital alliances, especially among Rajulu. Marriages in general are within caste groups but mostly observed outside the village among all the caste groups. Among Rajulu, possession of arable land and education are considered important for boys for finalising a marriage alliance. Usually in marriage, dowry is paid to groom's party, both in cash and kind (agricultural land, gold ornaments, etc.). Earlier, marital distance among Rajulu was around 30-50 km and the networks of relations were also limited. The amount of dowry was determined on the basis of ownership of property, family background and education.

Recently, it has been observed that the marital distance of both wife givers and wife takers has increased. Between 1994 and 1998 about 9 Rajulu girls were married out of which 4 were married to employers 3 to agriculturists and one to a businessman from other villages. The distance of giving their daughter in marriage ranged between 40-100 Kms. During the same period about 18 boys got married and brides were within a distance of 100-150 Kms., and most of them studied up to intermediate (75%) and graduation (25%). The amount of dowry given as part of marriage has also substantially increased.

The marriage expenses have increased multifold (about 3 times). During marriages money is spent lavishly. This spending also depends on their economic standing. The newly emerging class among Rajulu are trying to have relations with Rajulu in places such as Bhimavaram, Narasapuram, etc., in West Godavari district, who are economically better off. They now invite political leaders in the local area and notable *peddamanushulu* of all caste groups in the village during marriage functions in their families.

Earlier among Settibalija about half of the marriages were between cross cousins and close relatives. The marital distance has been largely within a vicinity of less than 40 Kms. The spending in the marriages was very limited. At present, not much difference is observed except for the fact that there has been a slight increase in the amount of dowry and marriage expenses.

As compared to Rajulu and other upper caste groups, preferential marriages were not high (20-30%) among the Scheduled Caste groups (both Mala and Madiga). The marital distance by and large was with in the vicinity of less than 25 km. Dowry and marriage expenses are slightly higher now than in the past.

Education

Generally education is considered as an indicator for human development. As discussed in Chapter-2, Godilanka has a high school and two primary schools. Students after completion of primary education join high school located in the central part of the village, for further studies. In addition to the residents of the village, children from the neighbouring villages also join the high school. Most of the students attending the high school belong to Scheduled Castes. However, there are a few from affluent sections sending their children to convents located at Komaragiripatnam and Amalapuram. This was the trend observed before the introduction of aquaculture.

In 1995 many small and marginal farmers have disposed off their cattle and land under paddy cultivation has come down. Hence the requirement of young boys to look after cattle has declined. This has led the Schedule Caste parents to send their children to school. The increased earnings due to aqua farming furthered the class and caste consciousness among Rajulu who shifted their children from the village school to convent schools located in neighbouring villages. The establishment of a convent school in Godi in 1996 run by a Raju has also helped the process of shifting their children to convent schools. Now about fourteen students are attending the convent school in Godi and twenty-two in Komaragiripatnam. These schools have transport facility. Though convent education is costlier Rajulu prefer to send their children to these schools as they have status implication.

Investment

Earlier surplus appropriated from agriculture and coconut garden was reinvested in land, ornaments, education of children, etc. In addition to ornaments, possession of cable TV and LPG including sending children to convent schools has now become points of discussion among women and for men it is automobiles, generator sets, etc.

Earlier among Rajulu, many were in debt trap. This was due to frequent crop failures in agriculture because of vagaries of nature. This to a certain extent was also due to growing costs of agricultural inputs, increasing demand for consumption loan to be advanced to labour, lower returns in agriculture and increasing household expenses. Besides these, children's education, lavish spending during life-cycle crises, rituals, festivals and marriages had always been adding to their debt. It is observed that about 65% of the small and marginal farmers, among Rajulu, were in debt ranging between 15,000 to 35,000 to the local moneylender. Approximately 45% of the medium farmers were indebted to an extent of Rs. 25,000 to 60,000 to PACS or moneylender and 20% of large and big farmers had a credit of over 50,000 to 1.5 lakh to bank, PACS or to the moneylender. After the aqua farming, all the farmers had cleared their earlier debts.

Electricity

In 1993, about 35 households from among the big, large and some of the medium landowner of the Rajulu caste had electricity and the source of lighting for others was kerosene lamps. Hardly any one from Settibalija caste had electricity either in Vasarlagudem or Vasamsettipalem. Among the Scheduled Caste, in Penumalapeta and Linepet, except for two teachers and four government employees, none had electricity. Among the others, one Golla household, who is a medium farmer, had electricity. At present, barring six households among Rajulu, all others have electricity among them.

Most of the new electricity connections (65%) were obtained in 1994 and the rest in the following years. In a similar fashion most of the Settibalija obtained electricity connections during 1995 in Vasamsettipalem and in Vasarlagudem in1998. About 20 households among Scheduled Castes had obtained electricity. The Pathuru Scheduled Caste members do not have electricity because electric poles have to be laid in the adjacent coconut garden for which the landowner is not consenting. Among the Scheduled Castes, Madiga do not have electricity.

Acquisition of other Assets

Acquisition of assets like television, automobiles, LPG, diesel power generators and other home appliances by different caste households indicate the impact of aquaculture on the living standards of the people. Possession of these assets indicates their improved economic status and social mobility.

Earlier only 5-6 households among Rajulu had television sets. After the establishment of a satellite television (locally, cable TV) in 1994 a medium aquafarmer had bought a colour television. The establishment of cable TV had induced many to possess a television. Subsequently, many farmers purchased same model TV sets. In the village, presently, about thirty farmers have procured colour television sets. Earlier only large and big farmers could afford them but now even marginal aquafarmers have them. Among Rajulu two aquafarmers have installed individual dish antenna. Besides, one group from Munsubgari *kutumbam* collectively installed a dish antenna to watch TV channels of their choice. A similar trend is also observed in case of refrigerators. Apart from Rajulu, about five Settibalija households have black and white TV sets, of which two are portable sets and eight households among the Scheduled Caste households, except for one, possess black and white TV sets, of which four are portable sets. The SC households bought the TV sets after 1995 only. This clearly indicates an improvement in the living standards of different caste groups, including the SCs.

Prior to aquaculture, large and big farmers exclusively possessed automobiles for transport. In Godilanka, in 1993, there were four to five two-wheelers and two cars. Almost all farmers owned bicycles, which were used to commute between house and fields, and sometimes to nearby villages. After aqua farming, many aquafarmers purchased two-wheelers. Every year, about three to four vehicles are purchased by aquafarmers. There are, presently, about 20 scooters and seven mopeds in Godilanka. Three aquafarmers have bought new cars in 1997. Possession of an automobile for aquafarmer has become a status symbol. They use these vehicles to visit their aqua farms and to frequent Amalapuram town. Among others, a Settibalija farmer and an employee now own mopeds and three Mala caste members also possess mopeds.

Earlier, the residue of coconut was widely used as fuel for cooking as it was abundantly available and was used by all, with an exception of four households who had LPG connection. LPG is one among the many modern items, which is observed in Godilanka after the introduction of aquaculture. The exposure and availability of LPG in the open market has made many housewives to opt for it as they regard it as a comfortable fuel for a kitchen. Presently, in Godilanka, there are about 50 LPG connections. However, almost all households still use firewood extensively for cooking, in spite of having a LPG connection. LPG is used only to make tea or coffee and is more a status symbol than a fuel for cooking medium.

In Godilanka there are frequent and long power-cuts daily. This, most of the time occurs just after the dusk and sometimes extends even during nights. This was construed as an inconvenience by landowners during summer. For a deliverance from this, in 1995, a group of Rajulu from *Rudraraju* lineage belonging to Munsabgari *kutumbam* had installed a diesel generator to generate power during power breakdowns. Cables were laid to all the members of the group. Each member had contributed equally. This installation was placed in the house of a person who would operate the generator, take care of its maintenance and to look after the account. The same group later installed a dish antenna on a collective basis. They are directly or indirectly affiliated to a regional political party. Location of their house in close proximity proved to be an advantage to this group. They are considered as role model for the other Rajulu caste households.

In a similar fashion to Munsabgari *kutumbam*, in 1997, another *Rudraraju* group belonging to large and medium category of cultivators and aquafarmers installed another generator in the more or less same procedure. This group is affiliated to a national political party. In 1998, one more group from *Rudraraju* and *Alluri* lineages, who are mostly medium aqua farmers, jointly has installed another power generator. President of PACS, who belongs to *Alluri* lineage, belongs to this group.

Nature of Interaction between the Landowner and Labourers

The physical movements of the various categories of people in Godilanka have changed to a large extent. Prior to aquaculture most of the supervisory *sonthamanishis* were either from Rajulu or Settibalija caste. Within the village the Settibalija caste members had greater accessibility to their Rajulu landowner's house and they could move freely within the house. They also used to act as messengers of the landowner to convey messages to other landowners. The movement of Settibalija caste members in the central part of the village was relatively higher over the other lower caste members. In the present context, barring a few Settibalijas, many have switched over to *gutta* or *paalu vyavasayam*, thus minimising their interaction with Rajulu landowners. From the point of view of Settibalija the movement from *paleru* or *sonthamanishi* to *gutta* or *paalu vyavasayam* is an upward social mobility. Thus the reduction in interaction with Rajulu landowners is not viewed as a lowering of status.

The movement of Schedule Caste members, who are employed as *sonthamanishi* and *paleru* with a Rajulu landowner, was restricted to the fields and taking care of cattle only. Except on occasions like unloading of paddy and coconut and during customary payments, Rajulu do not welcome them to visit their houses. Whenever they visit the landowners' houses they stay near the entrance of the house.

In the present context, Scheduled Caste members are engaged as agricultural/aquaworkers as their labour is more important. The SC aqua worker has to collect the feed from the Raju's house, keep the owner informed about the prawn condition in the tank, call the owner when a technician visits the pond, etc. This has made them visit the central part of the village more frequently, which is now widely accepted by the Rajulu. More importantly, during ordinary days and also during harvest days, he brings tea and on some occasions meals to his master. This is also the case with agriculture.

In the earlier situation, the place of interaction between the Settibalija and Scheduled Caste members was the place of work only. Settibalija *sonthamanishis* used to engage Mala or Madiga as agricultural labourers. However, it was rare for a Mala or Madiga to visit Vasarlagudem or Vasamsettipalem to meet a Settibalija *sonthamanishi*, except in times of need or crisis. Presently, since the Scheduled Caste labourers are in direct contact with Rajulu landowners they no longer needed the patronage of a Settibalija.

It is important to note that as a result of aquaculture two significant things have happened in Godilanka with regard to interactions among and between Rajulu, Settibalija and the SCs. While Settibalija have moved over to *gutta* or *paalu vyavasayam* their movement to Rajulu houses has come down substantially, this social space is now filled by the SC *sonthamanishis* or *palerus*. On the other hand, since the Settibalija no longer have any control over the labour of the SCs the interaction between them has become quite minimal.

The mobility of different groups of people to different other villages as well as others to Godilanka have also increased. Earlier, landowners used to go to Amalapuram, the nearest town, for procurement of agricultural inputs. Visiting relatives on social occasions was limited, as their social networks were more closed than they are today. The aqua situation has widened their networks. The feed supply companies conduct workshops and exhibitions at Amalapuram once or twice in a year and the aquafarmers from different villages attend them. This provides an opportunity for the Rajulu caste members from different villages to interact, thus provides a basis for widening their social networks. Further, visiting relatives along with their families have increased considerably. Possession of automobiles has further added to their interactions outside the village. Pleasure trips to distant places by peer aquafarmers after a harvest is now observed.

Settibalija having turned to *gutta* or *paalu vyavasayam*, now visit frequently Amalapuram for procuring agricultural inputs. There is also an increase in their visits to relatives' places and also attending social functions. An opposite trend is observed with the Scheduled Caste groups, who are employed as aqua workers. The aqua workers have to monitor the pond during the crop season and this has restricted them from visiting relatives outside Godilanka. However, they visit their relatives during the lean season.

Bargaining Capacity of Labourers

Prior to the introduction of aquaculture in Godilanka, the magnitude of employment of labour was high in agriculture and coconut gardens. These labourers, by and large, belonged to Settibalija, Mala and Madiga caste groups. After aquaculture in Godilanka, some labourers have adapted themselves as aquaworkers. Some of the Rajulu caste members, who became aqua farmers, leased-out their agricultural lands to earlier Settibalija *sonthamanishis* for *gutta vyavasayam*. This had increased the demand for agricultural labour, who are mostly from the Scheduled Castes, to work as aquaworker and also as *sonthamanishi* and *paleru* for them. The increase in demand for labour and withdrawal of labour of Settibalija has effectively increased the dependence of Rajulu on the SC labourers. As a result, the bargaining capacity of the labourers with the landowners, both in agriculture and aquaculture farms, has increased leading to higher wages.

Role and Status of Woman

The status of a woman is related to her father as well as husband's status. Tradition and the values in the contemporary society also have a bearing on her status. As their families move up in the social ladder women are withdrawn from the workforce outside the family and thus, are restricted to household work. Her work gets limited only to kitchen, begetting children and looking after the requirements of both children as well as husband. This always results in increased dependency on men. This can be observed mostly in the upper and middle ranking castes. Lower caste women however, contribute to the family income as they continue to be part of labour force. In all the caste groups, especially middle and lower castes, with the accumulation of wealth/ assets, there is a tendency to withdraw women form the work force making them more dependent on their counterparts (Siva Prasad, 1987: 240-4).

MN Srinivas sums up the status of Indian women as "when unmarried she is dependent on her father, after her marriage, on her husband, and when widowed, on her son... Such dependence is more acute among the better-off and the higher castes where women stay inside their homes. Immurement resulted in lessened knowledge of the outside world, and contributed to women's increased dependence on men, further fuelling the theory of perpetual dependence, women's activities became confined to the kitchen, and to the production and rearing of children" (1984: 25-6). This is quite true in the case of land-owning Rajulu women in Godilanka. These women do not work in the fields and are confined to household work only.

Among Settibalija, women from medium and small landowner categories do not work outside their households. However, most of the women from marginal landowners, tenants and landless labourers not only work on their farms but also work as agricultural labourers for others. Most of the Settibalija caste members are now in *gutta/paalu vyavasayam* and their women now work only on family farms. Some of them are now involved in dairying business. The other caste groups like Mangali, Kapu and Golla women do not work outside their homes, as was the case in the past. Bulk of women labour force still comes from Mala and Madiga castes.

Schism and Continuity: Seven-Generation Principle

All the members belonging to Rudraraju lineage among Rajulu are classificatory brothers. These members observe pollution relating to the life-cycle crises, viz., birth, puberty, menstruation and death. Since this lineage group has a large population they found it difficult to observe these restrictions. They frequently affected the observance of festivals, rituals, marriages, etc. Many a time marriages and other social functions were postponed due to death of a lineage member in the village. In order to overcome this, the lineage elders, who are also the *pedda-manushulu*, prior to the emergence of aquaculture, proposed to perform a ritual of 'breaking pots' for severing the bond of relationship among the members of lineage beyond seventh generation.

In 1996 the elders of Godilanka arranged for priests from West Godavari and Guntur. On an appointed day in May they had performed a *Yagna*, a ritual to propitiate Gods and ancestral spirits, and on the final day they broke the pots. This act symbolises the death of an individual in Hindu tradition[3], indicating spiritual separation of members. Following this ritual, the lineage of Rudraraju was divided into four *kutumbams* (families). On this was based on the fact that one of the ancestor had four sons and each son and his descendants symbolising a *kutumbam*. The four *kutumbams* are Markinda, Lacchi Raju (referred as Munsabgari *kutumbam* (as the Munsif was from this family), Gogulraju, and Tammiraju. Among these four *kutumbams*, Munsabgari *kutumbam* (Lacchi Raju) is regarded as the dominant section of the Rudraraju lineage, as they are both economically as well as numerically powerful. In contrast, Markinda is the lowest among the four *kutumbams*, as they are economically weak as compared to the others.

The relationships among the members of a *kutumbam* are noticed to be stronger than among the members of Rudraraju lineage as a whole. Here the point to be noted is that the extent of the relationship in the lineage has been reduced. The members of a *kutumbam* now consider themselves to be closer than the others in the lineage. Pollution is observed only at this level now. They participate in all social gatherings and have a greater concern to each other. They help each other financially and otherwise. They act as a corporate body in the village. It is to be noted that they are categorical and express themselves as belonging to the same family in discourses within and outside the village. They feel obligated to uphold the status of the *kutumbam* members and express their cohesiveness in public discourses. The *kutumbam* has now become important in *gutta vyavasayam* and in joint aqua farming. The members of a *kutumbam* are not barred from having affiliation with any political party.

Annually there is a fair held at the Gogulamma temple after Ugadi. It is customary for each *kutumbam* to perform *pooja* and offer sacrifices to the deity. The deity is taken in a procession in the village to facilitate villagers to offer their reverence to the idol of the goddess. This process earlier involved nominal expenditure, which was borne by the Rudraraju lineage members. However, after the division of this lineage into four *kutumbams*, each *kutumbam* started spending lavishly by illuminating the temple, using fireworks during the procession, arranging drummers, etc., on their allotted days of worship to display their status. Contrary to the situation prior to aquaculture, younger members are taking more interest in these activities to project the image of their *kutumbam*. The newly acquired wealth further seems to be strengthening the religiosity among the people, as it has become the medium through which they express their status elevation.

Settibalija now perform Sriramanavami with more enthusiasm. Since the last two years they are organising films shows for three days. These expenses are met from the contributions made by individuals for this occasion and part of it is also met by their caste *Sangham*. In addition to this, a few big landowners from Rajulu also contribute to this.

Among the Scheduled Castes, Malas celebrate Ambedkar Jayanthi[4] in Linepet and Penumalapeta separately. During this occasion, since 1997, they have been arranging film shows at both these places for which they contribute among themselves with few donations from landowners.

Ritual Differentiation

All the activities of ritual importance by landowners in Godilanka were venerated to Gogulamma, the village deity. There is an annual fair immediately after Ugadi. The cult of Lord Ayyappa, whose temple is located in Guruvayur in Kerala, is spreading in this region. Many from the upper castes have adopted it. After the introduction of aquaculture in Godilanka some Rajulu, especially those involved in aquafarming, started practising it. During the period of its observance one has to wear black dress, should treat everyone with respect and address him as 'swami', sleep on the floor, walk bare foot, cook food for himself, should take only vegetarian diet, should not smoke or consume liquor, forbidden to have sexual relations, etc. This has to be observed for 40 days, after which one has to visit the temple in Kerala. Popularly this is referred to as *'mala vesukovadam'* (wearing the ritual garland). After completing this ritual process, the devotees return to normal life. This is a type of cleansing of self. Every year about 10-15 members from Rajulu are wearing Ayyappa mala. This ritual performance by individuals relatively enhances their status among their kin and caste members and also in the village.

The cult of Bhavani (Goddess Kanakadurga), situated at Vijayawada is observed among Settibalija, both in Godilanka and in the region. Here the worshipper wears red dress and bangles. The worshippers are referred to as 'Bhavani'. The process is almost the same as in Ayyappa cult. After completion of the ritual period, while they set out to visit the temple all their caste members gather and touch the feet of Bhavani and give their offerings (money) to be submitted to the temple on their behalf. Every year about 8-10 members among Settibalija caste wear Bhavani. It is important to note that in Godilanka while Rajulu adopted Ayyappa worship (cult) Settibalija adopted Bhavani. This is a new ritual process noticed in Godilanka after the emergence of aquaculture. The lower caste groups resort to Bhavani, instead of Ayyappa cult, mainly due to economic considerations, as the devotees have to travel to Kerala, which involves higher expenditure than going to Vijayawada as in case of Bhavani. Thus, Ayyappa cult is considered superior to Bhavani cult. This is used as a means to elevate the status of individuals, who acquired wealth after aquaculture, among their relatives.

R. Nilakanta Raju is engaged in bringing LPG cylinuers on a moped from Amalapuram and deliver them to consumers. He got into this occupation from 1996 when many Rajulu have taken LPG connections for cooking. Earlier he was working as supervisory *sonthamanishi*. After aquaculture he switched over to this occupation. To raise his status among his relatives and villagers he had adopted Ayyappa cult. During this period he was not asked to do any demeaning jobs and was respected. On the day of leaving to visit the temple in Kerala, people, especially women, touched his feet as he was identified with divinity.

Alluri Pera Raju, 32 years, is a tenant aquafarmer. Earlier he was working as supervisory *sonthamanishi* with the same landowner with whom both his father as well as his grandfather worked. In 1995, due to aquaculture he was displaced from *paalu vyavasayam*. He along with his two brothers had leased-in 1.7 acres of aqua tank. After a good crop they have renovate their traditional house. All the three brothers have adopted Ayyappa cult in 1997. Most of the devotees of Ayyappa during this year stayed in the Panchayat office. This 'community of worshippers' develop a close personal rapport with each other and their families.

From the above cases it can be understood that some Rajulu, especially from the lower economic strata, to enhance their personal status in their caste and village, after improving their economic standing, use the cult of Ayyappa. Thus they use religion and ritual to legitimise their individual achievements in order to move up in the social ladder. The process is similar in case of Bhavani cult among Settibalija.

Informal Groups

In the village social organisation there exist a few informal groups of close kin members. These groups vary based on age, locality and interest within the village. Thus the village, for instance, consists of groups of boys, who generally belong to the same locality who move together, have recreation, meet in the evenings, etc., and discuss matters concerning them. These groups within the village overlap, interchanging members on different occasions. The core members of the group tend to remain over time. As these younger members of the village grow older, they tend to be accepted into older groups. These informal interest groups are found in all caste groups and they are related to their work and interest. Godwin (1972:57) observed a similar kind of informal group in sub-urban Bombay, where the principles were based on of sex, age, locality and interest principles.

After aquafarming in Godilanka, informal groups have increased in number. There are about 6-8 such informal groups in the village based on kinship among Rajulu, who are engaged in aquaculture. They help each other in their aqua enterprise in the village. Among the others, each caste group has on an average four such informal interest groups. These informal groups become the basis of interaction among individuals and these get extended even to the arena of recreation.

Most of the youth among Rajulu used to play volleyball during the lean agricultural season. They play both in the morning and evening. These youth have now become busy as many of them are involved in aquaculture. Since aquaculture required greater attention as it involves large investments, now they find no time for sports and games. There are a few youngsters in the village, who are educated in the towns, are more interested in playing cricket and are now involved in aquaculture. They play cricket near their aquaculture tanks occasionally. The television also has strong impact on the individuals. Recently cricket matches were organised between neighbouring villages and even a cricket tournament was also organised in the neighbouring village of Komaragiripatnam.

Gambling (playing cards) and cockfight are observed in Godilanka. Cockfights were more frequent among Rajulu in the village earlier. They consider it as a synonymous with their caste. Due to the multiplicity of informal interest groups and conflicts among them, the incidents of cockfights were reported to police. As a result, many were fined and some of them had to pay huge sums of bribe to the police. Consequently, now, the frequency of cockfights has come down. However, cockfights are still organised during marriages, festivals, parties, and other such occasions. The host bribes the concerned officials to avoid unpleasantness. During cockfight all caste groups in the village gather and gamble.

This can be clearly noticed in the marriage of Narayana Raju son of PACS President. Narayana Raju was a computer engineer and was married to a girl from west Godavari. In his marriage for three consecutive days cockfights were arranged. PACS President had sent an invitation to the people, including his relatives, in the neighbouring villages. On all these days he had made arrangements for their hospitality. On all these days there were cockfights and on an average, there were about 20 cockfights. Many of the villagers had also betting during the cockfight among themselves. Thousand of rupees were exchanged during this occasion on bets. The PACS President bribed the police and also used his familiarity with the local MLA for a smooth conductance of the occasion. This occasion was remembered for days in the village. This had enhanced his status and prestige among his relatives as well as in the village.

One more occasion in the village, which happened recently, also upholds the fact that lavish spending and arrangement of big parties enhances once prestige among the kin members and relatives. R. Ramakrishna Raju was a small farmer and he had been practising aqua farming since three years. A party was arranged by him on the occasion of his grandson's '*uyala*' (cradle) ceremony, a formal occasion when the child was placed in *uyala*. After the ceremony a grand dinner was arranged with various non-vegetarian dishes. All his relatives played cards in batches, based on their age and status, after the dinner. Liquor was provided according to their taste. Thus the party continued all through the night and attendants were arranged to supply them liquor and cigarettes. It is customary for Rajulu to play cards on occasions like parties, dinners, marriages, etc. These types of activities are commonly held in the houses of big landlords. In order to project his enhanced economic status after aqua farming, he had arranged for such a party. Such a thing was unthinkable for persons like Ramakrishna Raju under an agricultural enterprise. This occasion was used to consolidate the status of his *kutumbam* (Markinda), which had emerged as an informal power block in the village social organisation.

Decision-making Authority

In the agrarian context, among Rajulu, the entire economic power and decision-making authority was vested in the hands of the elders in the family. The eldest person, be it father or eldest son, decided on all the concerned matters. The sons or younger members were involved only in supervising the fields and had to obey his orders. Similarly, he maintained relations with the residents of the village, both his caste members as well as clients.

In the context of aqua farming, the economic power had slowly percolated down to the aquafarmers who largely belonged to the younger generation. However, the decision making relating to life-cycle rituals, marriage alliances, and other issues relating to their caste people, inter-caste relations and important decisions in the Panchayat are still vested with the older generation.

Among the Settibalija and the Scheduled Castes the decision-making is still in the hands of elders in the family. An important feature among them is that after marriage the sons start living separately and lead an independent life. However, they are consulted and their views are taken into consideration regarding matters relating to their families.

Thus the economic power and decision-making authority which was vested in the elders has now slowly moved into the hands of younger members, where the economic control switched its hands from older members to younger ones under aquaculture among land-owning Rajulu. In case of others, the authority still continues to be in the hands of older members.

NOTES

1. Traditionally, Uppara caste members are engaged in earth-digging work. They are also employed in de-silting canals, wells, etc.

2. It is Telugu New Year day, which usually falls in the month of March.
3. In general, a pot is broken before the pyre by the dead man's eldest son, to set his soul free from this world, thus, paving the way to the world of spirits (Heaven) in the midst of recitation of hymns (*mantra*) by the priests at the graveyard.
4. The birth anniversary of Dr. Babasaheb Ambedkar, the architect of Indian Constitution and the most revered leader of the Dalits, i.e., the deprived sections.

Chapter 6

Conclusion

In the earlier chapters we have discussed the role of aquaculture in fashioning the agrarian relations. The relations that have emerged in prawn culture are basically rooted in traditional agrarian relations and, in turn, these have their ramifications on the agrarian relations. Agrarian relations are viewed by some scholars as economic relations and are considered important in the production system of the villagers' social system. Others looked at them as essentially socio-cultural relations. The agrarian system is part of caste/class stratification with the upper caste/classes being the landowners and the middle caste/classes cultivating it and the labourers from lower caste/classes working as wage labourers. Land tenure, organisation of agriculture and the socio-economic conditions, which govern them, are regarded as constituent parts of the agrarian social structure.

The major contributor of change in the agrarian system and agrarian relations, as viewed by scholars, is economic development. With the exception of a few anthropological and sociological studies, studies on agrarian relations and agrarian structure were largely done from the point of view of economics. The studies by anthropologists and sociologists attempted to analyse the agrarian structure and relations from the point of view of traditions, values, stratification and other sociological issues. Their emphasis was to analyse the social/cultural change that has been taking place in the rural areas due to various development programmes undertaken by the Government as part of planned change. Further changes in the socio-economic and political spheres also have had their bearing on the agrarian social structure. These changes were viewed either as positional or structural. In fact, their perceptions were influenced by the kind of theoretical or ideological stance taken by them. The contributing factors for these changes as viewed by them are increased irrigation facilities, land reforms, new technology, commercialisation of agriculture and green revolution.

Some Marxist scholars characterised the production system in India chiefly as semi-feudal and semi-colonial. In aquaculture, as in agriculture, the relations of production are mainly between the aqua-owner and the labourers. Though aquaculture is an export-oriented enterprise, the relations between the owners and labourers tend to be pre-capitalistic with monetary payments. Unlike agriculture, there is no scope for payments in kind in aquaculture. Since the prawns are produced for the international market, they are beyond the reach of the common people, including the aquafarmers. The wages are calculated on the basis of the prevailing wage rates in agriculture. However, since there are no additional benefits that accrue to the labourers, unlike in agriculture, they receive a slightly higher wages.

The aqua-owners do not mind giving higher wages because aquaculture requires greater and careful monitoring and for this they need trust-worthy persons. Since they get higher margins of profit, they pay the labourers higher wages. The new category of aqua workers are not considered as attached labour by landowners, though the operating principles like consumption loan, wage structure follow largely the same conditions. They are excluded from the traditional payment. It is important to note that the aqua workers are mostly drawn from the agrarian categories of *sonthamanishi* and *paleru*, who worked with the same landowner earlier. To get an active cooperation from the labourers, who were earlier *sonthamanishi* and *paleru*, in place of customary payments they are given higher wages and consumption loans. The higher payments are *in lieu* of customary payments and the wage rates are arrived at after some negotiations. If the aqua workers continue to work both on aqua farms and agricultural fields of their owners, they receive both higher wages and customary payments.

The economic system of Godilanka, the study village, which was earlier predominated by agriculture, is now a combination of agriculture and aquaculture. The agricultural lands adjoining *Kouski* were converted to aqua ponds as these ponds had access to backwaters. The aquafarmers were the earlier agricultural landowners. The farmers who have switched over to the new lucrative aquaculture were entrepreneurial and had taken risk. They belonged to the higher strata and the earliest practitioners were progressive farmers, mainly the educated youth, who were quick to respond to this money spinning new enterprise.

Switching over to aquaculture, however, needed a quick adoption to new technology, rearing of prawn and risk-taking attitude. Thus the younger people from the dominant landowner caste, who have these abilities, have taken up this new enterprise. They found it more interesting as it did not involve hard physical work and the requirement of labour is also quite low. Thus the problems of labour recruitment and management are quite minimal in aquaculture. Largely it needed more monitoring and surveillance than agriculture and the returns are quite lucrative.

In the class of tenants some are owner-tenants and few do not possess own land but lease-in lands. Among the landless agricultural labourers, there are three distinct groups—the *paleru*, *sonthamanishi* and 'private', where the type of contract and the mode of payment differentiate them in agriculture and aquaculture.

In order to understand the changing agrarian relations in Godilanka due to aqua (prawn) farming, one has to comprehend the relationship between caste and class. Caste is an endogamous hierarchical group and to a certain extent a closed system. On the contrary, class is generally viewed as less rigid and is based on ownership of property, land and income. There are different economic classes in every caste and *vice versa*. But, there lies a cardinal difference between people of different castes in a class and different classes in a caste. Persons or families belonging to a similar class but from different castes are not equal in terms of access to resources or political power (Siva Prasad, 1987: 360-63).

Class is a construct, which largely depends on the economic position. It is more a conceptual category but is associated and identified with the status, i.e., on the ownership of landholdings and occupations. Each of these entities is not mutually exclusive but to a certain extent influence each other. In this context, it is important to study and understand the operating mechanisms of caste and class in the changing agrarian relations in Godilanka after the introduction of aquaculture. The basis of political power lies in the caste/class combine, as is clearly demonstrated in our study.

The top in the class structure of a caste is always at an advantage to receive the benefits of a new innovation/enterprise. The first to receive benefits in a new enterprise will always be from the upper crust of the dominant caste/classes. Invariably, the dominant castes are from the upper castes. Even in the past, the upper/dominant caste/classes preponderated in education, government services, agriculture, and, for that matter, any sphere of human development.

The study dealt with both caste and class as part of the agrarian social structure in Godilanka. The study mainly is oriented to analyse the categories of caste and class involved in a production system, their relations and interrelations. The factors such as the availability of and access to resources, and the individual attitude to venture into the new enterprise prompted the landowners to adopt aquaculture. The important aspects that helped the small and marginal farmers and leased-in farmers to take up aquaculture are kinship, caste and other primordial factors. These were also responsible for them to take up joint farming in agriculture. The control and access to resources, which have their roots in caste and class background of persons or families, gives them optimum advantage.

The life in a village usually does not differentiate caste and class into distinct entities but are always merged into the caste identity. The position of an individual in a caste group determines his access to resources and power. It was observed that those who lease-in land are predominantly from the upper caste, which possesses most of the arable land; the tenants belonged to middle rank castes; and the agricultural labourers invariably hailed from the lower castes.

The production process, to an extent, has changed from agriculture to aquaculture, while the ownership, more or less, continues to be in the same individuals/households from the same caste groups. This can not to be construed as structural change. For any change to be considered as structural change there has to be fundamental changes in nature of ownership of land and the relationships of production. The ownership of land is to a large extent in the hands of earlier landowners from Rajulu caste. Thus, in the absence of any change in the ownership of means of production, the changes in the relations of production would only be adaptive rather than structural. The change, which has occurred due to aquaculture, only reiterates the existing structure with some marginal realignments between the owners and workers. This confirms the theoretical proposition held in this thesis. Our study vindicates this point of view very clearly as can be noticed in the chapters 4 and 5. The agrarian relations under the impact of aquaculture are undergoing a change. As a result, they are redefining most of the earlier relationships and in this process there is continuity with the past structure of relations.

In aquaculture family and kinship continues to play an important role. The landowners involved in shared aquafarming are usually relatives, either consanguineal or affinal. Their participation may be on an equal basis or in proportion to their investment. Among the *paalu* tenants in agriculture, the labour exchange takes place amongst the close relatives.

In paddy cultivation, part of the produce was, and is, marketed and in case of aquaculture the whole production is for export. In spite of the changes in production, land-labour relations continue to operate on the earlier patterns. In this regard, it is important to note that theoretically when there is a change in the means of production the relations of production also change. However, the change in the mode of production if takes place without affecting the means of production, the relations of production relatively remain unchanged. This is quite adequately demonstrated in the 4th chapter.

The change in the production pattern from agriculture to aquaculture has brought a new type of tenancy, *gutta vyavasayam*, into existence. Earlier, the *sonthamanishi* was given a parcel of land for *paalu vyavasayam* and the inputs were supplied by the landowners. Presently, some of the earlier *sonthamanishis* from the middle and upper castes started leasing-in agricultural land for *gutta vyavasayam*. However, the others still lease-in land under *paalu vyavasayam*.

In aquaculture the nature of work and working patterns has changed to a large extent. There was almost a displacement of women labour as they had no designated work, so eventually they had diversified to different occupation like dairying working as house maids or servant maids. This had set-up new set of bondages both for men and women. The wage structure has increased for all categories of labourers. The incomes had increased the economic disparities among landowner, tenant and landless labourers. This is in conformity with the changes that could be observed in agrarian relations after the green revolution, raising crops for the market, development of irrigation networks and other developmental programmes. The income and profits have increased considerably to aqua farmers and *gutta* farmers, while only a little rise in wages can be observed for the workers who are seen in the lower rung of the production system.

In the occupation structure, which comprises of farmer, tenant and labourer, a new category of aqua worker came into existence. Based on the terms and conditions of work and mode of payment, aqua worker can be placed alongside of *sonthamanishi*, but certainly above *paleru*. He is neither totally free labour like 'private' or totally attached like *paleru* or *sonthamanishi*. He can be considered a type of attached labour wherein he is given consumption loan because of the responsibility bestowed on him. The associated occupations that have come up in the village are loose diesel vending, seed business, sale of lime and dolomite, nets and accessories, etc. Since these occupations require investment, these are restricted to the upper and middle castes. Among the landless labourers, those who work as private labourers continue to enjoy higher status over aqua worker, *sonthamanishi*, and *paleru*, in that order.

A caste differentiation is noticeable in the leasing-out process. Rajulu tenants lease-in land exclusively for *gutta vyavasayam*. The SCs lease-in agricultural land for *paalu vyavasayam* from landowners and this is linked to providing a *sonthamanishi* to the landowner. The Settibalija tenants lease-in land under both *gutta* or *paalu vyavasayam*. The earlier *sonthamanishis* from Settibalija caste are given land on lease by the landowners without any obligation of providing a *sonthamanishi*. Mere position of an individual in upper caste will not be the only criteria to have access to tenancy. A person's economic and political position in caste/class situation plays a significant role in this regard.

After the emergence of aquaculture caste *sanghams* have become rallying points for the members of the caste group. The *peddamanushulu* of the caste *sanghams*, who are generally from the upper strata, are more or less institutionalised authority structure and deal with the disputes and problems of their caste members, both within and outside the caste.

The agrarian relations in India were viewed as caste and class relations. The salient feature of these relations is mutual dependence between landowners and the workers. These relations are viewed as inter-caste relations and were considered under *jajmani* relations or patron-client relations. These relations are regarded as a sort of 'safety first' (Scott, 1987: 306) or insurance against natural calamities such as droughts, floods, epidemics and food crisis (Lipton, 1969:306). In the traditional context, *jajmani* payments were made even when there were crop failures (Bose and Jodha, 1965). These involve risk avoidance and sharing with fellow members of the village.

The developments that have taken place in agriculture after the green revolution have brought about changes in the *jajmani* system. Moreover, it is affected by many factors such as increase in population, labour shortages, migration, the decline of traditional occupations and increased employment opportunities. The relations that are found to be important for *jajmani* system are caste, kinship, locality, and affiliation to the village structure.

In agriculture, the labourers were paid either in cash or kind. Even today this continues to be the prevalent mode of payment for the labourers working in agriculture. The by-products in agriculture such as paddy straw are usually traded off for wages along with the grains. In aquaculture, which is export oriented, there are no by-products to be traded off. Consequently, the labourers are paid wages in cash. The labour relations are tending to be contractual but these relations are largely between the same set of people with, more or less, similar payments. It is essential to note that monetisation in aquaculture is a result of the nature of production and the aqua workers, who were mostly the earlier *sonthamanishis*, have adapted to the new situation. In this process of transition, a sort of continuity is noticeable.

Caste continues to be an important factor in determining one's place in the political structure of the village. The dominant caste controlled almost all the important centres of power in the village. With the coming of highly profitable aquaculture, some of the aqua farmers became quite rich. These new rich have become the contenders and started competing with the traditional leadership. They started developing connections/ networks outside the village and have built linkages with political parties. For the legitimisation of their achievements they frequently take an active role in the political activities of the village. The new rich arranged lavish parties on different occasions, in order to get recognition and to enhance their status among their relatives and in the village.

It is pertinent to note that with enhanced wages and a higher consumption loan the minimum necessities of the labourers are met. Presently, the aqua pond owners and labourers have a semblance of mutual understanding as aquaculture fetches both of them more money. The labourers do not have to toil hard using their physical labour, as their work on aqua ponds is more of monitoring.

Women from upper caste/classes feel that aquaculture did not make any difference to them, while it did affect adversely the Women from the other caste groups, as many of them were displaced from work. The members presently involved in *gutta vyavasayam* are withdrawing their women from working in the fields and restrict them to household work. However, there are a few who allow them to work in their own or leased-in fields. Women from the landless households contribute substantially to the family income. The displaced women have not only taken to dairying but also go for work in the neighbouring villages during peak agricultural season. There are a few who have even migrated to gulf. Thus, aquaculture had a significant impact on the women labour potential.

With the emergence of aquaculture, contrary to agriculture, the interactions between the landowners and labourers have increased as a result of increased dependence of aqua owners on the lower caste aqua workers. Since the money involved is much higher, the aqua owners need faithful and trustworthy persons and thus the earlier *sonthamanishis* are preferred as aqua workers.

The shaukars and commission agents continue to operate in the agrarian situation. Besides the above informal sources, aqua owners avail formal credit from the PACS in Godilanka. While the landowners/ aqua owners resort to the above sources for credit, the attached labourers such as *paleru*, *sonthamanishi* and aqua worker depend on the land/ aqua owners to receive consumption loan.

The Green Revolution has already weakened the traditional relations and customary payments for the service castes like vadrangi, Chakali and Mangali. The younger members from the service castes have established their own shops and they do not wish to enter into *jajmani* type of payment. However, the service castes continue to have patron-client relations with big and large farmers.

The introduction of aquaculture brought about increased circulation of money. This also had an impact on the value system. It is observed in our study that the preferences of landowners have changed from coconut and agriculture to aquaculture. As a result, large portion of agricultural land has been leased out by landowners. The newly acquired wealth is translated into status by way of acquiring new amenities and status enhancing things.

Relatively younger members by and large, control aquaculture and in the decision-making process, be it in the family, caste or village, their views are taken into consideration. With the accumulation of wealth, there is an intense political activity and lobbying among the aqua owners, which sometimes borders on factionalism. After the emergence of aquaculture, one can notice an increasing tendency among the aqua owners challenging the existing leadership. This, to a certain extent, is weakening the integrity and solidarity of political relations within the caste.

The nature and relations within the family structure have shown a degree of change. Division of property in a family continues to be only after all the children in a family get married. Aquaculture involves high risk, huge investment and needs greater cooperation among the members of the family and close kin members. This is clearly observed among the siblings from the undivided land owning families engaged in joint aqua farming. They share the resources among themselves, though they continue to live a nuclear life style. It is also worth mentioning that aquaculture is fostering jointness among landowners who are economically weak, and at the same time distinguishes from the better off who exhibit their strength and power being individual distinctly.

The status of an individual or a family was estimated in terms of the attached labourers working, the agricultural devices one possess such as tractors, power tillers, etc. After aquaculture, the demand for manpower has come down. The new symbols of status now are automobiles, LPG and modern kitchen appliances, possession of a colour TV with cable connection, sending their children to convent schools, etc.

The inequality in the ownership of land and resources among the landowners has resulted in differential access to benefits. The upper stratum of the landowners has a better access to facilities such as land and credit over the others. Owner-Labour relations are never static and they always adapt themselves to the changing situations with out affecting their basic relationship. The bondages between them can be comprehended in a dynamic context of changing relations. Worker attachment is a form of unfreedom and is a vital aspect in power relations as the owners control and extract cheaper labour. A kind of differentiation is observed among the farmers, particularly the small and marginal aqua farmers. They had quantum jump in their income after aquaculture. This has led them to purchase land in Godilanka or in other places where they have their kin members.

The present form of aquaculture and the relations of production reflect the past. In the processes of shifting to the new production system continuity with the past exists. New relations are always derived from the existing structure. Most of the times, they are adaptive changes and cannot be considered as structural changes. Earlier, the relations between the landowner and worker were on some kind of traditional arrangements. Now the relations between the aqua pond owner and aqua worker are mostly contracted. It will be misleading to presume that monetisation has basically altered the owner-labourer relationship under the new dispensation. The traditional bondages and structures mould the labourers' relationships with the pond owners to a large extent.

Based on our study, the following observations are made that are useful for policy and environmental protection. It also would help in reducing the tensions between the aqua farmers and paddy and coconut farmers. When there is a coexistence of paddy fields adjacent to aqua farms, there is a decrease in crop yields and this was found to be creating tension and conflict among the villagers. As suggested in the policy precautions have to be taken to dig a seepage canal round the tank, which collects the saline water and drains it out. Then the coexistence of both paddy and aquaculture becomes viable and cause little inconvenience to the adjacent paddy yields. This may reduce the conflicts and tensions between the farmers and the aqua farmers, which arise on account of this.

The changes that had been observed in the agrarian relations in Godilanka hold good, and are relevant, to any village in the region or elsewhere, which is under the impact of market forces as a result of liberalisation and globalisation. There is a need for studies of similar nature to study the impact of liberalisation and globalisation on the traditional economies where the ownership of the new enterprises lies still with the traditional rich. Further, there is need to understand the impact of these different types of capital intensive enterprises on the relationships among different categories of producers and workers as well as the surrounding physical environment.

References

Agarwal, Bina (1983), *Mechanisation in Indian Agriculture: An Analytical Study based on Punjab*, New Delhi: Allied Publishers.

Agarwal, Bina (1984), Rural Women and High Yielding Varieties in Rice Technology, in *Economic and Political Weekly,* March 31, pp. A 39-52.

Aggarwal, Pratap C. (1971), Impact of Green Revolution on Landless Labourers, in *Economic and Political Weekly*, November 20.

Ahluwalia, MS (1977), Rural Poverty and Agricultural Performance in India, in *Journal of Development Studies*, 13: 3 pp. 298-323.

Alagarswami, K (1991), Ensuring Sustainable Growth, in *The Hindu Survey of Indian Agriculture*, pp. 222-225.

Alagarswami, K (1995), Environment Issues of Shrimp Farming- Indian Experience, A paper presented at *Indaqua 95: Exposition of Indian Aquaculture*, January 27-30.

Alexander, KC (1973), Emerging Farmer-labour relations in Kuttanad, in *Economic and Political Weekly,* August 25, pp 1551-60.

Alexander, KC (1978), Changing Labour Cultivator Relations in Kerala, in *Social Action*, Vol. 28:3 July-September.

Alexander, KC (1979), Bonded Labour System, in *Man in India*, 59(2): 153-170.

Artherya, Venkatesh B, Djurfeldt and S Lindberg (1990), *Barriers Broken, Production Relations and Agrarian Change in Tamil Nadu*, New Delhi: Sage Publications.

Aziz, Abdul (1979), Irrigation and Economic Change in Karnataka Village, in *Studies in Rural development* (ed.) G. Thmmaiah Allahabad: G. Chugh Publications.

Aziz, Abdul (1980), Organising Agricultural Labourers in India: A Proposal. Kolkata: Minerva Associates.

Bailey, FG (1958), *Caste and Economic Frontier: A Village in Highland Orissa,* London: Oxford University Press.

Baird, Ian (1993), The Environmental and Social Costs of Developing Coastal Shrimp Aquaculture in Asia.

Banaji, Jaurus (1973), Mode of Production in Indian Agriculture, in *Economic and Political Weekly*, April 07, 1973 pp. 679-83.

Bandopadhyay, D (1979), A Tale from Telengana, in Das, Arvind (ed.) *Agrarian Relations in India*, New Delhi: Manohar Publications.

Bardhan, P (1980), Labour Employment and Wages in Agriculture: Results of Survey in West Bengal 1979, in *Economic and Political Weekly*, XV: 45-6 pp. 1943-1949.

Bardhan, P (1984), *Land and Rural Poverty*, Delhi: Oxford University Press.

Bardhan, P and A Rudra (1978), Interlinkages of Land, Labour and Credit Relations, An Analysis of Villages Survey Data in East India, in *Economic and Political Weekly*, February.

Bardhan, P and A Rudra (1980), Types of Labour Attachments in Agriculture– Results of a Survey in West Bengal, in *Economic and Political Weekly*, August 30.

Barraclough, S (1974), Politics First, in *Ceres* No. 41: 3, pp. 24-28.

Bayly, CA (1983), Rulers, Townsmen and Bazaars: North Indian Society in the Age of British Expansion, 1770-1870, Cambridge: Cambridge University Press.

Beidelman, TO (1959), *A Comparative analysis of the Jajmani System*, Monographs for the Association for Asian Studies VIII, New York.

Bell, Clive and TN Srinivasan (1973), Interlinked Transaction in Rural Markets: An Empirical Study of Andhra Pradesh, Bihar and Punjab, in *Oxford Bulletin of Economic and Statistics*, Vol. 51, pp. 73-83 and No. 329 March.

Benson, Janet (1996), A South Indian Jajmani System, in *Ethnology*, pp. 239-250.

Béteille, André (1969), *Caste, Class and Power—Changing Patterns of Stratification in a Tanjore Village*, Berkeley: California University Press.

Béteille, André (1972), The Traditions of Field Work, In *Economic and Political Weekly*, Special Number, August 1972.

Béteille, André (1974), *Six Essays in Comparative Sociology*, Delhi: Oxford University Press.

Béteille, André (1974), *Studies in Agrarian Social Structure*, Delhi: Oxford University Press.

Béteille, André (1991), The Reproduction of Inequality: Occupation, Caste and Family, in *Contributions to Indian Sociology*, 25 (1), 1991, pp. 3-28.

Bhalla, GS (1974), *Changing Agrarian Structure in India: A Study of the Impact of Green Revolution in Harayana*, Meerut: Meenakshi Prakashan.

Bhalla, GS and GK Chadha (1983), *The Green Revolution and the Small Peasant*, New Delhi: Concept.

Bhardwaj, K (1982), Production Conditions in Indian Agriculture, in Harriss, J (ed.), *Rural Development Theories of Peasant Economy and Agrarian Change*, London: Hutchinson.

Bhatt, S C (1998), *The Encyclopedia District Gazetteers of India Southern Zone*, Vol. 1, New Delhi: Gyan Publishing House.

Bhende, MJ (1986), Credit Markets in Rural South India, in *Economic and Political Weekly*, Vol. 21: 38 &39, pp. A 119-24.

Billings, MH (1970), Mechanisation and Rural Employment with some Implications for Rural Income Distribution, *Economic and Political Weekly*, Review of Agriculture June 27, 1970.

Blair, HW (1980), Rising Kulaks and Backward Classes in Bihar, in *Economic and Political Weekly*, XV: 2 pp. 64-74.

Blyn, G (1983), The Green Revolution Revisited, in *Economic Development and Cultural Change*, 31:4 pp. 705-25.

Bose, AB and NS Jodha (1965), The Jajmani System in a Desert Village, in *Man in India* Vol. 45:2 pp. 105-126.

Bose, PK (1984), *Classes in a Rural Society*, New Delhi: Ajanta Publications.

Bouglé, C (1971), *Essays on the Caste System*, Translated by D Pocock, Cambridge: Cambridge University Press.

Boyce, James K (1992), Of Coconut and Kings: The Political Economy of an Export Crop, in *Development and Change*, 23(4), pp. 1-25.

Brass, T (1986), Unfree Labour and Capitalist Restructuring in the Agrarian Sector: Peru and India, in *Journal of Peasant Studies*, Vol. 14(1).

Brass, T (1995), Unfree Labour and Agrarian Change: A Different View, in *Economic and Political Weekly*, April 1, pp. 697-99.

Breman, J (1974), *Patronage and Exploitation-Changing Agrarian Relations in South Gujarat, India*, Berkeley: University of California Press.

Byres, TJ (1972), The Dialectics of Indian Green Revolution, in *South Asia Review*, Vol. 5: 2.

Byres, TJ (1981), The New Technology, Class Formation and Class Action in the Indian Country side, in *Journal of Peasant Studies*, 8: 4 pp. 405-59.

Byres, TJ (1983), Sharecropping in Historical perspectives: A General Treatment, in *Journal of Peasant Studies,* Januvary/April pp.7-41.

Chandra, Bipin (1980), Peasantry and National Integration, in KN Pannikkar (ed.), *National and Left Movements in India*, New Delhi: Vikas Publishing House.

Chattopadhyay, P (1972a), On the Question of Mode of Production in Indian Agriculture: A Preliminary Note, in *Economic and Political Weekly*, VII 13: A39-46.

Chattopadhyay, P (1972b), Mode of Production in Indian Agriculture, in *Economic and Political Weekly,* VII: 53 pp. A185-192.

Cladwell, Bruce (1991), *The Jajmani System: An Investigation*, Delhi: Hindustan Publishing Corporation.

Coche, AG (1983), *Aquaculture and Development*, CIFA/ 83/5. FAO, Rome.

Cohn, BS (1955), The Changing Status of a Depressed Caste, in Marriott (ed.). *Village India*

Commander, S (1983), The Jajmani in North India: An Examination of Its Logic and Status Across Two Centuries, in *Modern Asian Studies*, 17,2 (1983), pp. 283-311.

Cooper, A (1983), Sharecroppers and Land lords in Bengal 1930-50: The Dependency Web and Its Implications, in *Journal of Peasant Studies* Vol.10: 2 & 3.

Csavas, Ime (1994), Advances and Limitations in Asian Shrimp Culture, in *Aqua International,* December pp. 8-27.

Dak, TM (1982), *Social Inequalities and Rural Development*, New Delhi: National.

Damle, CB (1989), Impact of Tenancy Legislation and Changing Agrarian Relations: A Case of Dakshina Kannada District, Karnataka, in *Social Scientist* (November-December), pp.83-97.

Damle, CB (1993), *Land Reforms and Changing Agrarian Relations*, Jaipur: Rawat Publications.

Das, Arvind N and V Nilakant (1979), *Agrarian Relations in India*, New Delhi: Manohar.

Dasgupta, Biplab (1977), India's Green Revolution, in *Economic and Political Weekly*, Annual Number, February 1977 pp. 241-60.

De Walt, BR, Philippe Vergne and Mark Hardin (1993), Shrimp Aquaculture Development and the Environment: People, Mangroves and Fisheries on the Gulf of Fonseca, Honduras, in *World Development,* Vol. 24:7 pp. 1193-1208.

Desai, AR (1971), *Peasant Struggles in India*, Delhi: Oxford University Press.

Dhanagare, B (1966), *Social Origins of Dictatorship and Democracy*, Middlesex: Penguin.

Dhanagare, D.N. (1987), Green Revolution and Social Inequalities in Rural India, in *Economic and Political Weekly,* May, Annual Number, pp. AN 137-44.

Dhar, H (1973), Agricultural Servitude—Bengal Presidency Around 1800, in *Economic and Political Weekly*, VII: 30 pp. 1349-1357.

Dharmaraj, J and Daisy Dharmaraj (1993), *Aquaculture: A Boon or Bane.*

Dharmaraj, J and Daisy Dharmaraj (1995), Peoples Movement Against Shrimp Industry—A Report, Madras: PREPARE.

Dube, SC (1955), *Indian Villages*, London: Routledge and Kegan Paul.

Dube, SC (1958), *India's Changing Villages: Human Factors in Community Development*, London: Routledge and Kegan Paul.

Dube, SC (1970), *Homo Hierarchies,* Chicago: Chicago University Press.

Dumont, L and Pococok (1958), Commented Summary of the First Part of Bouglé's, in *Contributions to Indian Sociology* 2:31-44.

Dumont, L and Pococok (ed.) (1957), *Contributions to Indian Sociology*, Paris: Mouton & Company.

Dutt, RP (1970), *India Today*, Kolkata: Manisha.

Duvvury, Nata (1985), Tobacco Trading and Farms of Market Organisation: A Case Study of Guntur District, in Raj, KN, Neeladri Bhattacharya, Sumit Guha and Saktipadhi (eds.), *Essays on the Commercialisation of Indian Agriculture*, Centre for Developmental Studies, Trivandrum and Delhi: Oxford University Press.

Duvvury, Nata (1986), Commercial Capital and Agrarian Relations: A Study of Guntur Tobacco Economy, in *Economic and Political Weekly,* July 26, pp. PE 46-57.

Epstein, TS (1962), *Economic Development and Social Change in South India*, Manchester: Manchester University Press.

Epstein, TS (1967), Product Efficiency and Customary Schemes of Rewards in Rural South India, in Firth, R (ed.), *Themes in Economic Anthropology*, ASA Monograph 6, London: Tavistock.

Epstein, TS (1978), *South India: Yesterday, Today and Tomorrow*, London: Macmillan.

Etienne, Gilbert (1968), *Studies in Indian Agriculture*, California: University of California Press.

FAO (1989), Aquaculture Production (1984-1987), FAO Fisheries Circular No. 815, Rev.1.

FAO (1989), *Review of Development in Aquaculture Committee on Fisheries*, 18th Session, Rome, 10-14 April, COFI/89/inf 5.

Farmer, BH (1977), *Green Revolution*, London: The Macmillan Press.

Flaherty, M Plate Vandergeest and Plate Miller (1999), Rice Paddy or Shrimp Pond: Tough Decision in Rural Thailand, in *World Development,* Vol. 27: 12 pp. 2045-2066.

Foster, GM (1973), *Traditional Societies and Technological Change*, Mumbai: Allied Publishers Private Limited.

Frankel, F (1971), *India's Green Revolution: Economic Gains and Political Costs*, Princeton: Princeton University Press.

Fuller, Christopher J (1989), Misconceiving the Grain Heap: A Critique of the Concept of the Indian Jajmani System, in Bloch, Maurice and Jonathan Parry (eds.) (1989), *Money and the Morality of Exchange*, Cambridge: Cambridge University Press.

Ganapathy, R (1989), Present Status of Prawn Farming in India, *Proceedings of the State level Seminar on Shrimp Farming in Tamil Nadu*, 13 May 1989, Fisheries College, Tuticorin.

George, Alex (1987), Social and Economic Aspects of Attached Labourers in Kuttanad Agriculture, in *Economic and Political Weekly,* December 26, pp. A 141-50.

Ghatak, S (1976), *Rural Money Markets in India*, Mumbai: Macmillan.

Ghurye, GS (1957), *Caste and Class in India*, Mumbai: Popular Book Depot.

Godwin, CJ (1972), *Change and Continuity*, Mumbai and New Delhi: Tata McGraw Hill Publishing Co. Ltd.

Goldman, Richard H and Lyn Squire (1982), Technical Change, Labour use and Income Distribution in Munda Irrigation Project, in *Economic Development and Culture Change*, Vol. 30: 4 July pp. 753-775.

Gorter, Pieter (1989), Canal Irrigation and Agrarian Transformation: The Case of Kerala, in *Economic and Political Weekly*, September 30,1989.

Gough, K (1955), The Social Structure of a Tanjore Village, in M. Marriott (ed.), *Village India*, Chicago: University of Chicago Press.

Gough, K (1960a), The Social Structure of a Tanjore Village, in MN Srinivas (ed.), *India's Village*, Mumbai: Asia.

Gough, K (1960b), Caste in a Tanjore Village, in Leach, E (ed.), *Aspects of Caste in South India and North-west Pakistan*, Cambridge: Cambridge University Press.

Gough, K (1974), *The Political Economy of Agrarian Change*, London: Macmillan.

Gough, K (1978), Changing Agrarain Relations in South Eastern India, in Current Anthropology Vol. 19:2.

Griffin, K (1972), *The Green Revolution: An Economic Analysis*, Geneva: United Nations Research Institute for Social Development.

Griglani, JM (1987), Coastal Aquaculture Potential and its Present Status of Prawn Farming in Andhra Pradesh, In Srivastava, UK, *et al.* (eds.), *Brackish Water Aquaculture Development in India.*

Gross, Jasper, David Burchh and Roy E Rickson (2000), Agri-Food Restructuring and Third World Transnational: Thailand, The CP Group and the Global Shrimp Industry, in *World Development*, Vol. 28:3 pp.513-530.

Guimãraes, Joaõ P De Campos (1989), Shrimp Culture and Market Incorporation: A Study of Shrimp Culture in Paddy Fields of Southwest Bangladesh, in *Development and Change*, 20(4), pp. 653-82.

Gulati, Leela (1984), Technological Change and Women's Work Participation and Demographic Behaviour: A Case study of Three Fishing Villages, in *Economic and Political Weekly*, Dec., 08, pp. 2089-94.

Gupta, AR (1984), *Caste Hierarchy and Social Change: A Study of Myth and Reality*, New Delhi: Jyothi Prakashan.

Gupta, SC (1963), *Agrarian Relations and Early Rule in India*, Mumbai: Asia Publishing House.

Gupta, SC (1969), Some Aspects of Indian Agriculture, in Desai, AR (ed.), *Rural Sociology in India.*

Hansra, BS and AN Shukla (1991), *Social, Economic and Political Implications of Green Revolution in India*, New Delhi: Concept Publishing Company.

Harper, EB (1959), Two Systems of Economic Exchange in Village India, in *American Anthropologist,* 61,1959 pp.760-778.

Harriss, B (1976), An Analysis f Rural Urban Transactions: An Indian and Sri Lankan Case Study, in Agriculture in the Peasant Sector of Sri Lanka, SWR de A Samarasingje (ed.), Paradeniya: Ceylon Studies Seminar Series.

Harriss, J (1961), Moneylending in the Village Economy of Malnad, in *Economic and Political Weekly*, February 4.

Harriss, J (1982), *Capitalism and Peasant Farming : Agrarian Structure and Ideology in Northern Tamilnadu*, Mumbai: Oxford University Publishers

Harriss, J (1991), The Green Revolution in North Arcot: Economic Trends, Household Mobility and the Politics of an Awkward Class, in Peter BR Hazell *et al* (eds.), *The Green Revolution Reconsidered: The Impact of High-yielding Rice Varieties in South India*, Delhi: Oxford University Press.

Hazell, Peter B *et al.* (1991), Economic Changes among Village Households, in Peter B, R Hazell *et al.* (eds.), *The Green Revolution Reconsidered.*

Herdt, W Robert (1987), A Retrospect View of Technological and Other Changes in Philippines Rice Farming 1965-1982, in *Economic Development and Culture Change*, Vol. 35: 2 January pp. 329-349.

Hewitt de Alacantra, C (1976), *Modernizing Mexican Agriculture*, Geneva: United Nations Research Institute for Social Development.

Hocart, AM (1950), *Caste: A Comparative Study*, London: Methuen.

ICLARM-GTZ (1991), The Context of Small-Scale Integrated Agriculture–Aquaculture Systems in Africa: A Case Study of Malawi, *ICLARM Studies Review*, 18, pp. 302.

ILO (1977), *Poverty and Landlessness in Rural Asia*, Geneva: ILO.

Infofish International (1990), Various Issues, Infofish, INFOPESCF, INFOPECHE and INFOSAMAL, the Regional Marketing Information and Advisory Services for Fish Production, Kaula Lumpur, Malaysia.

Ishwaran, K (1968), *Shivapur: A South Indian Village*, London: Routledge and Kegan Paul.

Issac, KO (1995), Aquaculture Pollution Much to do About Nothing, in *Aqua International*, August-September, pp. 6-13.

James, PSBR (1990), Problems and Prospects of Brackish Water Aquaculture in India, at *National Seminar on Recent Trends in Aquaculture,* 2-5 March 1990, CARE, Nagarjuna University.

Jetely, S (1977), Modernizing Indian Peasants—A study of Six Villages in Eastern U.P., India: Asian Educational Services.

Jha, Divakar (1967), Evaluation of Benefits of Irrigation, Tribeni Canal, Bihar, Mumbai: Orient Longman.

Jha, DN (1977), *Ancient India, An Introductory Outline*, New Delhi: People's Publishing House.

Jodhka, Surinder S (1995), Agrarian Changes, Unfreedom and Attached Labour, in *Economic and Political Weekly,* August 5-12, pp. 2011-13.

Jodhka, Surinder S (1995), *Debt Dependence and Agrarian Change*, Jaipur and New Delhi: Rawat Publications.

John, B Shaji (1996), The Environmental Impact of Aquaculture farms, in *Down to Earth*, April 15, pp. 93-5.

Joshi, PC (1967), Pre-Independent Thinking of Agrarian Policy, in *Economic and Political Weekly*, 11:8.

Joshi, PC (1969), Land Reforms in India, in AR Desai (ed.), *Rural Sociology in India*.

Joshi, PC (1970), Economic Development and Social Change in Traditional India, in *New Ideology*, Walter Rubens Felicitations Volume, German Academy of Social Sciences, Berlin: Central Institute of History.

Joshi, PC (1975), *Land Reforms in India: Trends and Prospects*, New Delhi: Allied Publishers.

Kapp, William K (1959), River Valley Project in India: Their Direct Effects, in *Economic Development and Cultural Change*, Vol. III: 1, October.

Karanth, GK (1987), New Technology and Traditional Rural Institutions: Case of Jajmani Relations in Karnataka, in *Economic and Political Weekly,* December 19, pp. 2217-24.

Karanth, GK (1995), *Change and Continuity in Agrarian Relations*, New Delhi: Concept Publishing Co.

Khan, W and RN Tripathy (1972), *Intensive Agriculture and Modern Inputs, Prospects of Small Farmers- WG District: Study in AP*, Hyderabad: NIRD.

Kishore, Vanguri (1995), Aquaculture and the Law, in *Eco News*, April-June pp. 4-6.

Kocherry, Thomas (1995), Intensive Coastal Shrimp Culture and Its Impact on Environment, a paper presented at *National Seminar on Current and Emerging Trends in Aquaculture and its Impact on Rural Development,* on 14-16 February.

Kolenda, P (1968), Religion, Caste and Family Structure: A Comparative Study of Indian Joint Family in Milton Singer and BS Cohn (ed.), *Structure and Change in Indian Society*, Chicago: Aldine Publishing Co., pp. 339-396.

Kolenda, P (1978), *Caste in Contemporary India: Beyond Organic Solidarity*, California: The Benjamin Cunnings Publishing Company.

Konar, Harekrishna (1977), *Agrarian Problems of India*, Kolkata: Ganashakh Printers.

Kori, Nijera (1996), Profit by Destruction, *International Workshop on Ecology, Politics and Violence of Shrimp Cultivation for Export*, Dhaka: Nijera Kori.

Korovkin, Tanya (1992), Peasants, Grapes and Corporations: The Growth of Contract Farming in Chilean Community, in *Journal of Peasant Studies*, January 1992 pp. 228-254.

Kotausky, G (1964), *Agrarian Reforms in India*, Delhi: Peoples Publishing House.

Krishanamurthy, J (1972), The Growth of Agriculture Labour in India—A Note, in *IESHR*, 9:3.

Krishna, Mohana K (1995), Prawn/Shrimp Culture, The Scourge of Coastal India, in *Eco News*, April-June pp. 4-6, also presented at *National Conference on Sustainable Agriculture,* April 5-6, 1995.

Krishnaji, N (1971), Wages of Agricultural Labour, in *Economic and Political Weekly*, VI: 38, pp. 148-151.

Kumar, D (1962), Caste and Landless in India, in *Comparative Studies of Society and History*, Vol. IV.

Kumar, D (1965), *Land and Caste in South India*, Cambridge: Cambridge University Press.

Kumari, S and K Rangarao (1993), Development and Change: A Study of Jajmani Relations, in Bhat, Chandrasekhar, Kadikar Laxminarayana and K Ranga Rao (eds.), *Sociology of Development and Change*, New Delhi: Orient Longman.

Kushro, AM (1973), *Economics of Land Reform and Farm Size in India*, Delhi: The Macmillan Company.

Kushro, AM (1973), How Green is the India's Green Revolution, in *Economic and Political Weekly,* December 29, pp. A.133-44.

Leach, ER (ed.) (1960), *Aspects of Caste in South India, Ceylon and North-west Pakistan*, London: Cambridge University Press.

Lenin, V (1967), *Development of Capitalism in Russia*, Moscow: Progress Publishers.

Lerche, Jens (1993), Dominant Castes, Rajas, Brahmins and Inter-caste Exchange Relations in Coastal Orissa: Behind the Facade of the Jajmani System, in *Contributions to Indian Sociology*, 27(2), pp. 237-66.

Lewis, O (1958), *Village Life in Northern India—Studies in a Delhi Village*, Urbana-Champaign: University of Illinois Press.

Liao, Chiu (1992), Aquaculture in Asia: Status Constraints, Strategies and Prospects, in Liao, Chiu, Chung-zen Shyu and Nai-Hsien Chao (eds.), *Aquaculture in Asia: Proceedings of the 1990 APO Symposium in Aquaculture*, Taiwan Fisheries Research Institute.

Liao, IC (1988), East Meets West, An Eastern Perspective of Aquaculture, in *Journal of World Aquaculture Society*, 19(2): 62-73.

Lipton, M (1969), The Theory of the Optimising Peasant, in *Journal of Development Studies*, Vol. 4.

Lipton, M (1982), Why Poor People Stay Poor, in Harris J (ed.), *Rural Development: Theories of Peasant Economy and Agrarian Change*, London: Hutchinson.

Lipton, M and R Longhurst (1989), *New Seeds and Poor People*, Baltimore: John Hopkins University Press.

Lukose, A (1991), *Labour Movements and Agrarian Relations*, Jaipur: Rawat Publications.

Majumdar, DN, MC Pradhan, C Sen and S Misra (1955), Intercaste Relations in Cohanakallan , a Village Near Lucknow, in *Eastern Anthropologist*, March- August 8:191-214,.

Mandalbaum, D (1955), Technology, Credit and Culture in a Niligiri Village, in McKim Marriott (ed.), *Village India*, Chicago: The University of Chicago Press.

Marriott, M (1960), *Social Structure and Change in a UP Village: Risk and Vulnerability*, Delhi: Oxford University Press.

Marriott, M (ed.) (1955), *Village India*, Chicago: University of California Press.

Marx, K (1954), *Capital* Vol. 4, Moscow: Progress Publishers.

Mathur, KS (1958), Caste and Occupation in a Malwa Village, in *Eastern Anthropologist*, 12 (1), 1958 pp. 47-61.

Mayer, AC (1960), *Caste and Kinship in Central India: A Village and its Region*, Berkeley and Los Angeles: University of California Press.

McGuire, Thomas, R. (1983), The Political Economy of Shrimp in the Gulf of California, in *Human Organization*, 42 (2), pp. 132-45

Mehta, RC and DS Sidhu (1971), A Study of Impact of Medium and Long-term Loans on Short-term Credit Needs of Small Farmers, in *Indian Journal of Agricultural Economics*, October-December.

Mehta, Sushila (1980), *A Study of Rural Sociology in India*, New Delhi: S Chand and Company.

Meillassoux, C (1973), Are There Castes in India, in *Economy and Society* Vol. 2: 1.

Mellor, JW (1976), *The New Economics of Growth*, New York: Cornell University Press.

Mencher, P. Joan (1975), Agricultural Labour Movements in their Socio-Political and Ecological Context: Tamil Nadu and Kerala in Balakrishnan Nair (ed.), *Culture and Society*, India: Thompson Press.

Mencher, P. Joan (1978), Agrarian Relations in Two Rice Regions of Kerala, in *Economic and Political Weekly*, Annual Number, February, pp. 349-66.

Mencher, P. Joan (1980), The Lessons and Non-lessons of Kerala: Agricultural Labourers and Poverty, in *Economic and Political Weekly*, XV: 42-43, pp. 1781-1802.

Michie, H. Barry (1973), Green Revolution: An Anthropological Perspective, in *Economic and Political Weekly*, Review of Agriculture, June.

Miller, Daniel (1975), *From Hierarchy to Stratification: Changing Pattern of Social Inequality in North Indian Village*, Delhi: Oxford University Publishers.

Miller, Daniel (1986), Exchange and Alienation in the Jajmani System, in *Journal of Anthropological Research*, Vol. 42:4 pp. 535-556.

Misra, GP and M Vivekand (1979), Impact of Canal Irrigation in Rural Areas (Case of Bellary District), in *Impact of Irrigation, Tank, Well, Canal*, New Delhi: Concept Publishing Company.

Mohammad, Noor (1988), *Caste and Primary Occupations*, New Delhi: Concept Publishing Company.

Mohanty, BB (1999), Agricultural Modernisation and Social Inequality: Case Study of Satara District, in *Economic and Political Weekly*, June 26, pp. A50-61.

Mohapatra, K (1996), *Exploitative Tenancy and Technological Backwardness in Indian Agriculture, A Case Study of Orissa*, Mumbai: Himalay Publishing House.

Morales, Edmundo (1986), Cocoa and cocaine Economy and Social Change, in *Economic Development and Culture Change*, Vol. 35:1 October,, pp. 143-161.

MPDEA (1991), *Handbook of Shrimp Farming*, Cochin: MPDEA.

Mukherjee, Ramakrishna (1957), *The Dynamic of a Rural Society*, Berlin: Akademic Verlag.

Mukherjee, Ramakrishna (1981), Realities of Agrarian Relations in India, in *Economic and Political Weekly,* January 24, pp. 109-16.

Mukhopadhyay, Rajatsubhra (1980), Role of Caste in Agrarian Relations, in *Social Scientist*, Vol. 8:9 April.

Mukul (1994), Aquaculture Boon: Who Pays?, in *Economic Political Weekly*, December 3 pp. 3075-78.

Muralidharan, CM and Shankar (1995), Shrimp Farming Boom-Effect on the Coastal Poor and their Environment, a paper presented on 5-6 April at National Workshop on *Sustainable Aquaculture* at Anna University.

Murthy, CS (1987a), Impact of Technological Changes on the Extent of Tenancy and on the Composition of Tenants: A Study on Two Delta Villages, in *Indian Journal of Agricultural Economics*, Vol. 42:3 (July-August).

Murthy, CS (1987b), Influence of Socio-Economic Status on Contractual Terms of Tenancy: A Study in Two Delta Villages of Andhra Pradesh, in *Economic and Political Weekly*, (Review of Agriculture), Vol. XXII: 39, September 26.

Murthy, KVR (1995), Shrimp Culture and the Environmental Issues, in Parthasarathy, G and JVH Dixitulu (eds.) (1995).

Naidu, V Chandersekara (1997), Agricultural Transformation and Poverty: Some Issues for Discussion, *Review of Development and Change*, Vol. II: 1 pp.122-33.

Namboodripad, EMS (1979), Caste Conflicts Vs Growing Unity of Popular Democratic Forces, in *Economic and Political Weekly* XIV (7-8), 329-336.

Naqvi, S (1975), Marx on Indian Pre-Asiatic Society, in Mathew Kurian (ed.), *India-State and Society*, Mumbai: Orient Longman.

Nayar, PKB (1976), Emerging Pattern of Agrarian Movement in India, in *Kerala Sociologist*, Vol. IV: 2, December.

NCAER (1974), *Credit Requirements for Agriculture*, New Delhi: National Council for Applied Economics and Research.

NIRD (1975), *Changing Agrarian Relations in India*, Hyderabad: National Institute of Rural Development.

Okhawa, Kazukh (964), The Role of Agriculture in Early Economic Development—A Study of Japanese case in Economic Development with Special reference to East Asia (ed.), L Berrily (1964).

Olsen, Wendy Kay (1996), *Rural Indian Social Relations*, Delhi: Oxford University Press.

Omvedt, Gail (1980), Caste, Agrarian Relations and Agrarian Conflicts, in *Sociological Bulletin*, 29(2), pp. 142-170.

Omvedt, Gail (ed.) (1982), *Land, Caste and Politics in Indian States*, Delhi: Delhi University Press.

Oomen, TK (1971), Green Revolution and Agrarian Conflict (Revolution by Production Function), in *Economic and Political Weekly*, VI: 26, pp. A99-103.

Oomen, TK (1989), Impact of the Green Revolution on the Weaker Section, in Sharma, ML and TM Dak (eds), *Green Revolution and Social Change*.

Opler, ME and RD Singh (1952), Economic, Political and Social Change in a Village in North-Central India, in *Human Organization* Vol. II: 2 pp.5-12.

Orenstein, H (1962), Exploitation or Function in the Interpretation of Jajamani, in *Southwestern Journal of Anthropology* 18:302-15.

Pandey, MP (1979), The Impact of Irrigation on Rural Development, A Case Study, New Delhi: Concept Publishing Company.

Parry, J (1979), *Caste and Kinship in Kangra*, New Delhi: Vikas Publishing House.

Parthasarathy, G and JVH Dixitulu (eds.) (1995), *Blue Revolution and Public Policy*, Hyderabad: Booklinks Corporation.

Parthasarathy, G. and V. Pothana (1983), Capitalism in Agriculture and Peasant Farming: Case Study of Godavari Delta, in *Economic and Political Weekly*, September 24, pp. A 93-97.

Parthasarathy, G (1970), *Agricultural Development and Small Farmers: A Study of Andhra Pradesh*, Delhi: Vikas.

Parthasarathy, G (1984), Growth Rates and Fluctuation of Agricultural Production: A District-wise analysis in AP, in *Economic and Political Weekly*, June 30, pp. A 74-84.

Parthasarathy, G (1991), HYV Technology: The Polarisation or Immiserisation Controversy, in *Economic and Political Weekly*, June 29, pp. A 69-77.

Parthasarathy, G and B Prasada Rao (1969), *Implementation of Land Reforms in Andhra Pradesh*, Kolkata: Scientific Book Agency.

Patel, SJ (1952), *Agricultural Labour, India and Pakistan*, Mumbai: Current Book House.

Patil, RG (1980), An Investigation into the Socio-economic conditions of GHOD Command Area (Maharashtra), Department of Agriculture Economics, MP, Krishividhyapeeta.

Patnaik, Utsa (1971a), Capitalist Development in Indian Agriculture—A Note, in *Economic and Political Weekly*, VI, No. 39, pp. A123-130.

Patnaik, Utsa (1971b), Capitalist Development in Indian Agriculture, in *Economic and Political Weekly* VI, pp. A190-194.

Patnaik, Utsa (1972a), On the Mode of Production in Indian Agriculture- I, in *Economic and Political Weekly*, VII 40, pp. A135-151.

Patnaik, Utsa (1972b), Development of Capitalism in Agriculture-II, in *Social Scientist*, Vol.1: 3.

Patnaik, Utsa (1973), Development of Capitalist Production in Agriculture, in *Social Scientist.*

Patnaik, Utsa (1990), *Agrarian Relations and Accumulation, The mode of Production Debate in India*, Mumbai: Oxford University Press.

Pearse, A (1980), *Seed of Plenty, Seeds of Want*, London: Oxford University Press.

Peterson, S (1982), Allocation of Aquaçultural Resources, in LJ Smith and S Peterson (eds.), *Aquaculture Development in Less Development Countries: Social, Economic and Political Problems*, Boulder: Westview Press.

Plateau, J. Ph, J. Murickan and E Delbar (1980), Rural Credit Market in Backward Areas—A Kerala Fishing Village, in *Economic and Political Weekly*, October (Special Number).

Plateau, J. Ph, J. Murickan and E Delbar (1981), *Interlinkage of Credit, Labour and Marketing Relations in Traditional Marine Fishing: The Case of Purakkad (Kerala)*, Delhi: Hindustan Publications.

Poffenberger, Mark and Mary S Zurbuchen (1980), The Economics of Village Bali, Three Perspectives, in Economic Development and Cultural Change, Vol.29:1 October pp.91-133.

Prasad, PH (1976), Poverty and Bondage, in *Economic and Political Weekly,* August.

Prasad, PH (1980), Rising Middle Peasantry in North India, in *Economic and Political Weekly*, XV: 5-7 pp. 215-219.

Prasada Rao (1995), A Study of Impact of Aquaculture on Environment in Coastal Andhra Pradesh, in Parthasarathy and JVH Dixitulu (eds.), 1995.

Pullin, RSV, H. Rosenthal and JL MaClean (eds.) (1993), *Environment and Aquaculture in Developing Countries*, ICLARM Conference Proceedings.

Radhakrishnan, P (1982), Land Reforms and Changes in Land System: Study of a Kerala Village, in *Economic and Political Weekly*, September 25, pp. A 107-119.

Radhakrishnan, P (1983), Land Reforms and Social Change: Study of a Kerala Village, in *Economic and Political Weekly*, December 24-31, pp. A143-50.

Raghava Rao, DV (1979), Modern Farm Technology: Sociological Aspects, in Singh, JP *et al.*, *New Seeds Adoption and Yield*, New Delhi: Sterling Publishers Pvt. Ltd.

Rajpurohit, AR and K Pilliar (1977), Agriculture, Irrigation Transfer and Its Socio-Economic Consequences on Mandya, Bangalore: Shiny Publishing Company.

Ramakrishna Rao, C Siva (1986), *Changing Agrarian Relations in India*, Secunderabad: M.Venkata Rangaiya Foundation.

Randive, BT (1979), Caste, Class and Property Relations, XIV 7-8 pp.337-348.

Rao, CHH (1970), Farm Size and Credit Policy, in *Economic and Political Weekly*, Review of Agriculture, December 26.

Rao, CHH (1975), *Technological Change and Distribution of Gains in Indian Agriculture*, Delhi: Macmillan.

Rao, GN (1977), Agrarian Relations in Coastal Andhra Under Early British Rule, in *Social Scientist*, 6 (1), pp. 19-29.

Rao, GN (1985), Transition from Subsistence to Commercialised Agriculture: A Study of Krishna District of Andhra, C 1850-1900, in *Economic and Political Weekly*, June 22-29, pp. A 60-9.

Rao, Krishna YV *et al.*, (1984), *Peasant Farming and Growth of Capitalism in Indian Agriculture*, Vijayawada: Vishalandra Publishing House.

Rao, MSA (1972), *Tradition, Rationality and Change*, Mumbai: Popular Prakashan.

Raychaudhuri, T (1982), The State and the Economy: The Mughal Empire: Non-agricultural Production: Mughal India Trade, in *Cambridge History of India*, Vol.1.

Reddy, A Ranga (1990), *Agricultural Development, Rural Credit and Problems of Its Recovery*, New Delhi: Mittal Publications.

Reddy, NS (1955), Functional Relations of Lohars in a North Indian Village, in *Eastern Anthropologist*, 8:129-40, March-August

Reddy, Sanjeeva (1995), Blue Revolution: Need for Progressive and Pro-active Public Policies in Parthasarathy and JVH Dixitulu (eds.), *Blue Revolution and Public Policy.*

Reddy, VN and CS Murthy (1978), Backward Caste and Tenancy: A Village Study, in *Economic and Political Weekly,* July 1, pp. 1061-76.

Reddy, VN and CS Murthy (1985), Pattern of Commercial Farming in Guntur District, in *Economic and Political Weekly,* December 21-28, pp. A 137-52.

Rosin, T (1987), *Land Reform and Agrarian Change, Study of Marwar Village from Raj to Swaraj*, Jaipur: Rawat Publications.

Roth, HD (1983), *Indian Moneylenders at Work*, New Delhi: Manohar.

Rubino, Micheal C and Richard W Stoffle (1990), Who will Control the Blue Revolution? Economic and Social Feasibility of Caribbean Crab Mariculture, in *Human Organisation*, 49:4, pp.386-94.

Ruddle, Kennth (1993), The Impact of Aquaculture Development on Socio-Economic Environment in Developing Countries: Towards a Paradigm for Assessment, in Pullin, RSV, H. Rosenthal and JL MaClean (eds.), *Environment and Aquaculture in Developing Countries*, ICLARM Conference Proceedings, 31 pp.20-41.

Rudra, A (1969), Big Farmer of Punjab, in *Economic and Political Weekly*, IV: 59, pp. A143-146.

Rudra, A (1970), In Search of Capitalist Farmer, in *Economic and Political Weekly*, V: 25, pp. 85-87.

Rudra, A (1971), Capitalist Development in Agriculture—A Reply, in *Economic and Political Weekly*, VI: 45, pp. 2291-92.

Rudra, A (1982), *Indian Agricultural Economics: Myths and Realties*, New Delhi: Allied Publishers.

Saberwal, Satish (ed.) (1972), *Beyond the Village, Sociological Explorations*, Simla: IIAS.

Sachchidananda (1972), Social Dimensions of Agriculture Development in Bihar (with IADP), Delhi: National.

Sachchidananda (1988), *Social Change in Village India*, New Delhi: Concept Publishing Company.

Sachchidananda and BB Mandal (1983), Agrarian Relations: A Case Study of Purnea in Bihar, in *Man in India*, 63 (1), pp. 3-20.

Sakthivel, M (1985), Shrimp Farming: A Boon or Bane to India, *ICLARM Newsletter*, 8(3): 9-10.

Santhanakrishnan, G (1985), Brackish Water Shrimp Culture in India—Present Status and Future Prospects, in Training *Manual on Brackish Water Shrimp Culture and Extension*, MPDEA, Cochin, pp.1-26.

Santhanakrishnan, G (1994), Changing Horizons in Indian Aquaculture Scenario, in *Aqua International*, January, pp. 13-4.

Santhanakrishnan, G (1995), Ecological Issues Vs. Sustained Development in Shrimp Culture, in *Aqua International*, May, pp.6-10.

Santhanakrishnan, G (1996), Violence of the Blue Revolution, in *Aqua International*, January, pp.38.

Saradamoni, K. (1982), Woman's Status in Changing Agrarian Relations: A Kerala Experience, in *Economic and Political Weekly*, January 30, pp. 155-62.

Sarap, K (1987), Transactions on Rural Credit Markets in Western Orissa, India, in *Journal of Peasant Studies*, Vol. 15: 1 October.

Sarap, K (1991), *Interlined Agrarian Markets in Rural India*, New Delhi: Sage.

Sastri, KA Nilaakant (1966), *A History of South India*, London: Oxford University Press.

Satyanarayana, G. (1992), *Changing Agrarian Structure and Labour Relations*, Jaipur: Rawat Publications.

Schneider, Todd (1991), Indian Shrimp Culture, *International Fisheries Report*, 91/18, March 14.

Schoer, Karl (1977), Agrarian Relations and the Development of the Forces of Production, in *Social Scientist*, Vol. 6: 3 October.

Scott, James (1976), *The Moral Economy of the Peasant: Rebellion and Subsistence in South-east Asia*, New Haven: Yale University Press.

Scott, James (1987), Peasant Moral Economy as a Subsistence Ethic, in Shanain. T (ed.), *Peasants and Peasant Societies: Selected Readings*, pp.304-310.

Sen, Bhowani (1962), *Evolution of Agrarian Relations in India*, New Delhi: Popular Publishing House.

Sengupta, Nirmal (1979), Caste as an Agrarian Phenomenon in Twentieth Century Bihar, in Das, Arvind (ed.), *Agrarian Relations in India*, New Delhi: Manohar Publications.

Shah, Ghanshyam (1977), *Protest Movements in Two States*, Delhi: Ajanta Publications.

Shah, SC and LR Singh (1970), The Impact of New Agricultural Technology on Rural Employment in the Punjab, in *Indian Journal of Agricultural Economics,* Vol. 25:33-39

Shah, SM (1969), Rural Class Structure in Gujrat, in AR Desai (ed.), *Agrarian Sociology in India*, Mumbai: Popular Prakashan.

Shanain, T (ed.), (1987), *Peasants and Peasant Societies: Selected Readings*, Oxford: Basil Blackwell.

Shankar, Kripa (1999), Agricultural Reforms: Rhetoric and Reality, in *Economic and Political Weekly,* November, 20, pp. 3264-65.

Sharma, ML and TM Dak (1989) (eds.), *Green Revolution and Social Change*, Delhi: Ajanta Publications.

Shergill, HS and Gurmail Singh (1995), Poverty in Rural Punjab: Trends Over Green Revolution Decades, in *Economic and Political Weekly*, Vol.30: 25 pp. A80-83.

Shiva, Vandana (1994), The Damaging Social and Environmental Impact of Aquaculture, in *Third World Resurgence,* No. 59, pp. 22-4.

Shivakumar, SS (1978), Aspects of Agrarian Economy in Tamilnadu: A Study of Two Villages, in *Economic and Political Weekly*, XIII:18, pp. 762-770; XIII:19 pp.812- 23; and XIII: 20, pp.846-51.

Shivakumar, SS and Chitra (1979), Class and Jati at Asthanapuram and Kanthapuram, in *Economic and Political Weekly,* XIV 6-7, pp. 263-286.

Siddiqui, Kalim (1999), New Technology and Process of Differentiation: Two Sugarcane Cultivating Villages in UP, in *Economic and Political Weekly*, December 25, pp. A139-152.

Silas, EG (1980), Status of Prawn Culture in India and Strategy for Its Future Development: *Proceedings, First National Symposium on Shrimp Farming*, 16-8 August 1978, Mumbai, pp. 209-18.

Silas, EG (1983), Prawn Culture and its Importance in National Development, *Proceedings, Summer Institute on Hatchery Production of Prawn Seeds and Culture of Marine Prawn*, 18 April-17 May 1983, CMFRI, Cochin.

Silvertsen, D (1963), *When Caste Barriers Fall*, London: Asllen & Unwin.

Singh, JP, KV Govindaraju, M Prahaladachar, S Giriappa, AR Rajapurohit, M Venkata Reddy, DV Raghava Rao (1979), *New Seeds Adoption and Yield*, New Delhi: Sterling Publishers Pvt. Ltd.

Singh, VB (1975), Village Community, in VB Singh (ed.), *Economic History of India 1857-1956*, Mumbai; Allied Publishers.

Singh, Yogendra (1969), Changing Power Structure in UP villages, in Desai, AR (ed.), *Rural Sociology in India.*

Singh, Yogendra (1969), Social Structure and Village Panchayats, in Desai, AR (ed.), *Rural Sociology in India.*

Siva Prasad, R (1987), *Social Mobility in Bangalore City*, Unpublished Ph.D. Thesis submitted to University of Mysore, Mysore.

Sohani, DK and RD Khandarkar (1970), Pattern of Income, Distribution, Savings and Expenditure In Rural Areas, in *Indian Journal of Agricultural Economics*, 23:133 July-September.

Somayajulu, VVN and K Mahanth (1984), Rural Transformation in Two Districts of AP, in *Social Change,* December, pp. 41-46.

Srinivas, MN (1955), The Social System of a Mysore Village, in M. Marriott (ed.), *Village India*, Chicago: University of California Press.

Srinivas, MN (1966), *Social Change in Modern India*, Mumbai: Allied Publishers.

Srinivas, MN (1976), *The Remembered Village*, London: University of California Press.

Srinivas, MN (1984), *Some Reflections on Dowry*, Delhi: Oxford University Press.

Srivastava, Ravi (1989a), Interlinked Modes of Exploitation in Indian Agriculture During Transition: A Case Study, in *Journal of Peasant Studies*, 16:4, pp.493-522.

Srivastava, Ravi (1989b), Tenancy Contracts During Transition: A Study Based on Fieldwork in Uttar Pradesh (India), in *Journal of Peasant Studies*, 16:3, April pp.334-395.

Srivastava, UK, B H Dholakai and S Vathsala (eds.) (1987), *Brackish Water Aquaculture Development in India*, New Delhi: Concept Publishing company.

Subba Rao, M (1995), Shrimp Culture in Andhra Pradesh, A Paper Submitted in the *Conference on Displacement Due to Aquaculture,* at TISS on 20-22 September.

Subba, TB (1984), Caste and Agrarian Structure: A Study in Hill Darjeeling and Sikkim, in *Man in India*, Vol. 64:1, pp. 59-68.

Sundararaj, V, M Devaraj and MJ Prince Jeyaseelam (1992), Prawn Farming in India, *in* Liao, Chiu *et al., Aquaculture in Asia: Proceedings of the 1990 APO Symposium in Aquaculture.*

Swain, Mamata (1999a), Agricultural Tenancy and Interlocked Transactions-1, Neoclassical and Marxist Approaches, in *Economic and Political Weekly*, September 11, pp. 2657-2666.

Swain, Mamata (1999b), Agricultural Tenancy and Interlocked Transactions-1, Neoclassical and Marxist Approaches, in *Economic and Political Weekly*, September 18, pp. 2752-2758.

Swamy, Dilip S (1988), Agricultural Tenancy in the 1970s, in *Indian Journal of Agricultural Economics,* Vol.43: 4, pp.555-568.

Tharakan, Abraham (1990), *Shrimp Production in India: A Review.*

The Imperial Gazetteer (1908), Vol. XII, Oxford: The Clarendon Press.

Theodorson, George A and AG Theodorson (1970), *A Modern Dictionary of Sociology*, London: Methuen & Co. Ltd.

Thorner, Alice (1982), Semi-feudalism or Capitalism: Contemporary Debate on Classes and Modes of Production in India, in *Economic and Political Weekly,* December 4, 1982, pp. 1961-68; December 11, 1982, pp. 1993-99; December 18, 1982, pp. 2061-66.

Thorner, D (1956), *Agrarian Prospect in India*, Delhi: Oxford University Press.

Thorner, D (1962), *Land and Labour in India*, Mumbai: Asia.

Unni, Raman (1975), Sources of Agricultural Labour in Kerala: Some Social Perspectives, in Balakrishna Nair (ed.), (1975), *Culture ad Society*, India: Thompson Press.

Upadhyaya, Carol Boyack (1988), The Farmer-Capitalists of Coastal AP, In *Economic and Political Weekly,* July 2, 1988, pp. 1366-82; and July 9, pp. 1376-1442.

US Embassy (1990), National Marine Fisheries Science.

US Embassy and Consulate Reports, Kolkata, December 26,1990.

US Embassy and Consulate Reports, Kolkata, July 12,1991.

US Embassy and Consulate Reports, Chennai, December 18,1991.

Varahalaraju, B (1980), *Sri Andhra Kshatriya Vamsa Ratnakaramu*, Vol. 1, Batanavalli: Sri Sharada Printing Press.

Varghese, TC (1970), *Agrarian Change and Economic Consequences*, Mumbai: Allied Publishers.

Vasudeva Rao, D (1987), *Rural Development Through Irrigation*, New Delhi: Ashish Publishing House.

Walinsky, LJ (ed.), (1977), *Agrarian Reforms As Unfinished Business: The Selected Papers of Wolf Ladejinsky*, New York: Oxford University Press.

Warriner, Doreen (1969), *Land Reform in Principle and Practice*, Oxford: Clarendon Press.

Weidner, Dennis and Roschberry (1992), World Shrimp Farming, *Proceedings of the Special Session on Shrimp Farming, World Aquaculture Society* Annual Meeting, May 21-25, pp. 1-21.

Wiser, WH (1969), *The Hindu Jajmani System*, Lucknow: Lucknow Publishing House.

Xaxa, Virginus (1988), Studies in Indian Agrarian Social Structure: A Survey of Trends and Perspectives, in *Eastern Anthropologist* 41: 1, pp.17-48.

Zaman, MK (1982), Change and Continuity in the Patron-client Relations Among Peasants of Bangladesh, in *Eastern Anthropologist*, 35 (4), pp. 271-84.

Index

PLATES

1. Sivalayam

2. Anjaneya temple

3. Rice Mill on the Central Part of the Village

4. Gram Panchayat Office

5. Village Tank near the Gram Panchayat Office

6. *DabuIllu* with Dish Antenna and a *Logi Illu*

7. Bantapani Group Digging Aquapond near PHC

8. Bantapani Group Cleaning the Bottom of the Pond

9. Progressive Aquafarmer Overseeing the Preparation of Ponds by Migrant Uppara Community

10. Aqua Worker Treating the Newly converted Pond from Coconut Gardens with Agricultural Lime

11. Filling the Aquafarm with Tubewell Water

12. Dying Coconut Tress: Environment Effect of Aquaculture

13. Temporary Huts of Migrant Uppara Community in a Coconut Garden